TRENDS AND DEVELOPMENTS OF MODERN
SUSTAINABLE BUILDINGS

GREEN
ARCHITECTURE IN CHINA

绿色建筑在中国
——现代节能建筑趋势与发展

高迪国际出版有限公司 编

王丹 刘宪瑶 李小童 陈曦 译

大连理工大学出版社

图书在版编目(CIP)数据

绿色建筑在中国：现代节能建筑趋势与发展：英汉
对照 / 高迪国际出版有限公司编；王丹等译. — 大连：
大连理工大学出版社, 2014.7
　　ISBN 978-7-5611-9150-7

　　Ⅰ.①绿… Ⅱ.①高… ②王… Ⅲ.①生态建筑—研
究—中国—英、汉 Ⅳ.①TU18

　　中国版本图书馆CIP数据核字（2014）第100248号

出版发行：大连理工大学出版社
　　　　　（地址：大连市软件园路80号　邮编：116023）
印　　刷：上海锦良印刷厂
幅面尺寸：240mm×320mm
印　　张：22
插　　页：4
出版时间：2014年9月第1版
印刷时间：2014年9月第1次印刷
责任编辑：初　蕾
责任校对：王丹丹
封面设计：高迪国际

ISBN 978-7-5611-9150-7
定　　价：348.00元

电　话：0411-84708842
传　真：0411-84701466
邮　购：0411-84708943
E-mail：designbookdutp@gmail.com
URL：http://www.dutp.cn

如有质量问题请联系出版中心：（0411）84709246　84709043

PREFACE I

Andrea Destefanis (left)

Co-Founder and Co-Chief
Architect, Kokaistudios

安德烈·德斯特法尼斯（图左）
Kokaistudios 建筑事务所联合创始人、
联合首席建筑师

The term "green" is one of the most widely used but not clearly defined terms in architecture today.

If we search for a textbook definition of "Green Architecture" it can be roughly summarized as a philosophy of architecture that advocates the use of renewable energy sources, the efficient use of water, the reuse and safety of building material sands the sitting of buildings with consideration of the surrounding environment.

In China we see several structural problems in the process of building that undermine the potential for the growth and meaningful spread of green architecture; the two most glaring being the speed of construction project developments and the low cost of construction. The combination of these two critical elements means that the number of truly "green" architectural projects is statistically minute and the vast majority of ordinary buildings do not even fulfill the most elementary building standards from a point of view of materials, energy conservation and insulation.

While we push for the inclusion of these technologies in our projects and in certain cases like our recently completed Shanghai K11 project, in which the extensive use of green technologies resulted in the awarding of a LEED Gold certificate, we don't think that focusing on strictly "green" architecture is the right answer for China.

We view "green" as one part of a more complex approach; that approach is one of "sustainable architecture". "Sustainable", in comparison to "green", is not simply an environmental concept. The environment is one part of a three-sectioned relationship that also includes social and economic components. In such, sustainability as an approach is far more complex than simply offering to save energy or reduce carbon emissions and we would argue that this more nuanced approach can deliver far greater benefits to China than simply focusing

在如今的建筑领域，"绿色"一词应用十分广泛，但含义却很模糊。

如果按照教科书上的解释，"绿色建筑"可以简单地定义为一种建筑理念，即主张使用可再生能源，有效利用水资源，建筑材料安全可靠且可以重复使用，建筑与周围环境和平共处。

在中国，在建造过程中我们会发现一些结构上的问题，它们破坏了绿色建筑的发展和绿色建筑深远意义的传播；其中最明显的两个问题就是建筑项目发展的速度过快和建筑成本过低。这两个关键问题的结合意味着真正"绿色"的建筑项目的数量很少。从建筑材料、能源保护和隔热能力方面看，绝大部分的普通建筑甚至连最基本的建筑标准都达不到。

我们争取将这些技术融入到我们的项目中，在某些案例，比如我们最近完成的上海 K11 项目中，应用了大量的绿色技术，因而获得了能源与环境设计先锋奖金奖的认证，我们认为，对于中国而言，只将重点放在绿色建筑上，并非明智之选。

我们将"绿色"看作是一种更加复杂的方法，是"可持续建筑"的一部分。"可持续"与"绿色"相比，不仅仅是一个环境概念。环境是三部分关系的其中之一，其他两部分是社会和经济。这样看来，可持续性这一方法比单纯地节约能源或减少碳排放量要复杂得多。我们认为这个更加细致的方法会为中国带来更多的益处，而不是仅仅局限于发展"绿色"建筑，因为可持续建筑可以大规模升级扩展。

on developing "green" buildings as it is massively scalable.

For this reason, Kokaistudios has focused since our first projects in China in the fields of renovation and adaptive reuse.

Let's start by considering the life cycle of a building, and the concept of "embodied energy". Embodied energy can be defined as the energy consumed by all of the processes associated with the production of a building ranging from the acquisition of natural resources to product delivery and this includes mining, manufacturing of materials and equipment, transport and administrative functions to demolition.

By reusing buildings, one retains their embodied energy making the project intrinsically much more environmentally sustainable than entirely new construction. If one couples the benefits of embodied energy with a well-designed retrofit emphasizing green design solutions then we have a very sustainable solution.

From an economical perspective there are the initial financial savings and returns to be made from adaptive reuse of existing buildings related to the "embodied energy" of the buildings; there are time savings to be made from the fact that there is no need for major demolition and re-starting the construction project from a foundation level; and there are additional potential benefits related to issues of existing buildings versus new government regulations.

Keeping and reusing historical buildings has long-term social benefits for the communities that value them and adaptive reuse, when done with thought and care, can restore and maintain the heritage significance of a building and help to ensure its survival. Rather than falling into disrepair through neglect or being demolished, heritage buildings that are sympathetically re-developed can continue to be used and appreciated by the local society.

由于这个原因，Kokaistudios 建筑事务所从在中国的第一个项目开始就专注于整修和改建的领域。

让我们从建筑生命周期和"蕴含能源"的概念开始谈起。蕴含能源可以这样理解：与建筑生产有关的，从采获自然资源到产品传送的所有步骤中消耗掉的能源，包括采矿、材料和设备制造、运输和行政拆迁。

通过对建筑的再利用，可以保持蕴含能源，使翻新的建筑本质上比全新的建筑更加具有环境可持续性。如果能秉承蕴含能源的理念，再加上注重绿色环保的精心整改设计，那么我们会得到一个可持续性很高的解决方案。

从经济角度看，秉承"蕴含能源"观念，我们可以通过节省现存建筑修缮的开支，来得到利益回报；不将建筑项目进行大规模推倒重建可以节省时间；我们还有可能从政府的有关现存建筑的新规定中获得潜在利益。

保护并重建具有历史意义的建筑，对公众而言具有深远的社会意义。对建筑进行认真仔细的修缮可以使建筑历史意义得以保留，并使其流芳百世。对富有历史意义的遗产建筑进行整修，避免其因受忽视或拆迁而遭受毁坏，这样不仅能延长建筑的寿命，还能获得当地社会的认可与感激。

PREFACE II

Alistair Lathe

Associate Director, Atkins
李仕达，阿特金斯副董事

From what began with small pioneering efforts in the late 20th century to develop buildings that would fundamentally use fewer resources, it was always the goal of sustainable architecture to bring about a transformation that would be of global significance. Today, we see green architecture coming to the forefront in China during a period of greater urbanisation than the world has ever witnessed.

Not only has sustainable design emerged on the scene, but a gradual maturation is now underway, leading to the application of established practices and new initiatives on an ever increasing scale. Thanks to a genuine and growing interest from the general public, and supported by a higher position on the political agenda, green buildings in China are firmly in demand.

This new interest comes with the expectation that sustainable architecture should be something to behold, in the same manner by which technological innovation has visibly transformed other aspects of society. While technology certainly features highly in its development, in truth many of the most effective green building methods are "passive" – understated or hidden from view altogether. Fundamental strategies such as high-performance glazing, low-consumption water fixtures and a well-insulated building envelope go thanklessly unnoticed.

The desire for a visual expression of sustainability has led to the trend of eye-catching features to make a green building's credentials more apparent. Solar panels, wind turbines and green walls can all contribute to reducing carbon footprint, but are too often employed to demonstrate a commitment to environmental action, more than helping achieve that purpose. Sadly, this has led to a degree of cynicism on the part of the informed professional community.

So, what kind of statement should a contemporary green building make?

The answer will surely reveal itself in the broader context in which architecture must be understood. Environmental considerations have always been fundamental to the design of buildings, alongside a myriad of other concerns and aspirations. What has changed is the expectation of a new aesthetic, addressing an issue which, though now critical, was always present.

In the face of our current necessity to transform how and what we build, a multi-faceted

从 20 世纪末期微小的开拓努力到使用更少的资源进行建造，一直以来，可持续性建筑的发展目标都是引领一场有国际影响力的变革。今天，中国正经历着世界上前所未有的城市化进程，在这个阶段，绿色建筑开始显露头角。

可持续性设计并非刚刚出现，而是已渐渐走向成熟，因此，惯用手段和新举措都得到了更广泛的应用。绿色建筑在中国之所以有如此稳固的需求，要归功于公众兴趣的热情高涨以及政府高层的支持。

人们对这一新兴领域充满兴趣是因为他们认为可持续性建筑应该和其他技术革新一样可以改变社会。毫无疑问技术在自身领域起到重要的作用，然而实际上，大部分高效的绿色建筑方法是"被动的"——被轻描淡写或被完全抹杀掉了。像高性能玻璃、节水装置和完全绝缘的建筑外壳等的重要性都被忽视了。

过于追求可持续性在视觉上的表现，导致了只注重视觉特性的趋势，因此使绿色建筑的认证更加流于表面。太阳能电池板、风力涡轮机和绿色墙壁的确可以减少碳排放量，但更多时候这些做法只是为了体现环保承诺，而并非真正能达到环保目的。可悲的是，这样的观点会引起见多识广的圈内专业人士的冷嘲热讽。

那么，当代的绿色建筑到底应该是怎样的？

必须将建筑业放在更广阔的背景下，那样答案就会不言自明了。环境因素与其他各种繁多的考量及意愿一起，一直都是建筑设计的根本。已经改变的是新的审美期望，以期解决一直存在着的，但最近变得迫切的问题。

面对当下，我们必须有所转变，建设什么？怎样建设？我们需要一个多方面的解决方法。与那些不那么显眼

approach is needed. Overt, recognizably "green" features which draw people's attention to the environmental impact of buildings do have a place, alongside less conspicuous but proven means and measures. Perhaps nowhere is this more pertinent than in China, where the development of today's cities – and the involvement of its citizens – will have implications around the globe.

With buildings making a major contribution to energy consumption (accounting for over 40% of energy use in some countries), it is the responsibility of architects and engineers to lay out the roadmap towards a low-carbon society. While it is encouraging to see a lot of ambition across the profession, many efforts are uncoordinated and remain superficial and ineffective. The key to meaningfully implement and deliver low carbon design is through an integrated, multi-disciplinary design approach.

At Atkins, we have the capability and depth to bring together a wide range of disciplines and skills, from the beginning to the end of the design process. On one hand it is the creative thinking of our people that shapes the approach we call "Carbon Critical Design". At the same time, the complexity of our built environment requires new and advanced tools to quantify, forecast and measure carbon efforts.

Atkins has developed a comprehensive suite of carbon tools which are designed to help make decisions on how best to reduce embodied and operational carbon. These tools reflect the full breadth of our multi-disciplinary capabilities from master planning to buildings, energy management to transport, and progressing from high-level options to detailed design choices. In addition to assessing carbon reductions, the long term pay-back period and thus return on investment can also be forecast. This is of particular interest to the private sector, which clearly has a large role to play in the global effort to tackle climate change and pave the way towards a true low-carbon society.

The projects in this publication reveal just how far the scale and ambition of green architecture has already come, in parallel with the intensifying growth of our cities and built environment. What will be interesting to see is not only how people respond to innovative green designs, but how we gradually adopt to the changes in lifestyle and urbanism that underlie this emerging architecture.

Stefan Abidin
Senior Associate Director, Atkins
林兆恒，阿特金斯高级副董事

但已有案例证实的方法手段相比，显然"绿色"这一特色标签的确占有一席之地，能够让人注意到建筑的环境影响。也许不会再有哪里能像在中国这样如此贴切，如今中国城市的发展和市民的参与度都会对全球产生影响。

随着建筑在能源消耗中比重的增加（在一些国家占能源使用量的40%以上），为低碳社会描绘发展蓝图就成了建筑师和工程师的责任。虽然许多努力与结果并不对等，仍然流于表面而且效率低下，但看到在这一职业中的许多雄心壮志仍然很令人振奋。确保低碳建筑设计实施的关键就是要有综合的、多学科的设计方案。

在阿特金斯，我们有能力和洞察力将贯穿设计过程始终的各种准则和技巧进行深度整合。一方面，是人们的创造性思维造就了我们所说的"低碳设计"。同时，建筑环境的复杂性要求新兴、先进的工具来量化、预测及测量碳排放量。

阿特金斯发展了一套综合碳处理工具，以此来判断怎样才是减少隐含碳和可用碳的最好方法。这些工具反映了我们全面的跨学科能力，从总体规划到楼体建造，从能源管理到运输，能够处理高端设置也能对微妙的设计做出判断选择。除了评估碳排放量的减少之外，还可以预测长期投资回报期和其投资回报量。私营企业对此喜闻乐见，毫无疑问，他们在全球致力于扭转气候变化的进程中扮演着举足轻重的角色，并为建立一个真正的低碳社会铺平道路。

从本书的案例可以看出，绿色建筑目前已达到相当的规模，具有远大的抱负。与城市和建筑环境的急剧增长并驾齐驱、相辅相成。我们关注的不仅是人们对绿色设计革新的反应，还有我们如何逐渐适应生活方式和城市化进程的变化，这些都推动着新兴建筑的发展。

CONTENTS

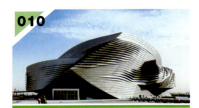

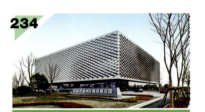

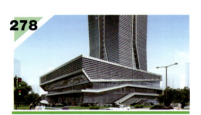

DALIAN INTERNATIONAL CONFERENCE CENTER

大连国际会议中心

SUSTAINABLE & GREEN FEATURES 绿色特征

- *Thermal energy of seawater with heat pumps for cooling in summer and heating in winter* ／ 利用海水热能和热泵，夏季制冷，冬季供暖

- *Low temperature systems for heating in combination with activation of the concrete core as thermal mass to keep the building on constant temperature* ／ 低温加热系统和混凝土芯蓄热相结合使建筑恒温

- *Large use of natural ventilation and daylight* ／ 大量使用自然通风和自然光

- *Energy production with solar energy panels* ／ 太阳能发电

ARCHITECT

COOP HIMMELB(L)AU, Wolf D. Prix / W. Dreibholz & Partner ZT GmbH

DESIGN PRINCIPAL

Wolf D. Prix

PROJECT PARTNER

Paul Kath (until 2010), Wolfgang Reicht

PROJECT ARCHITECT

Wolfgang Reicht

DESIGN ARCHITECT

Alexander Ott

DESIGN TEAM

Quirin Krumbholz, Eva Wolf, Victoria Coaloa

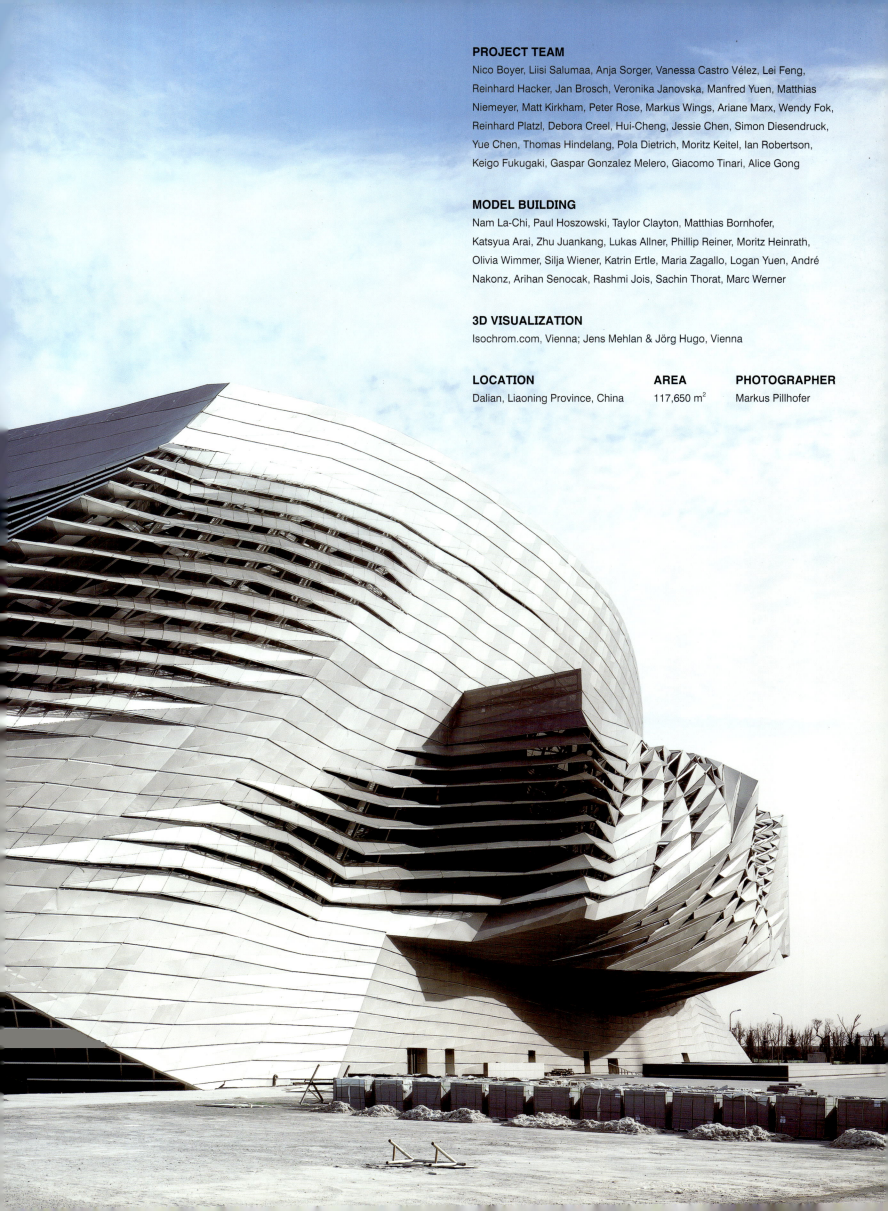

PROJECT TEAM

Nico Boyer, Liisi Salumaa, Anja Sorger, Vanessa Castro Vélez, Lei Feng, Reinhard Hacker, Jan Brosch, Veronika Janovska, Manfred Yuen, Matthias Niemeyer, Matt Kirkham, Peter Rose, Markus Wings, Ariane Marx, Wendy Fok, Reinhard Platzl, Debora Creel, Hui-Cheng, Jessie Chen, Simon Diesendruck, Yue Chen, Thomas Hindelang, Pola Dietrich, Moritz Keitel, Ian Robertson, Keigo Fukugaki, Gaspar Gonzalez Melero, Giacomo Tinari, Alice Gong

MODEL BUILDING

Nam La-Chi, Paul Hoszowski, Taylor Clayton, Matthias Bornhofer, Katsyua Arai, Zhu Juankang, Lukas Allner, Phillip Reiner, Moritz Heinrath, Olivia Wimmer, Silja Wiener, Katrin Ertle, Maria Zagallo, Logan Yuen, André Nakonz, Arihan Senocak, Rashmi Jois, Sachin Thorat, Marc Werner

3D VISUALIZATION

Isochrom.com, Vienna; Jens Mehlan & Jörg Hugo, Vienna

LOCATION	**AREA**	**PHOTOGRAPHER**
Dalian, Liaoning Province, China	117,650 m²	Markus Pillhofer

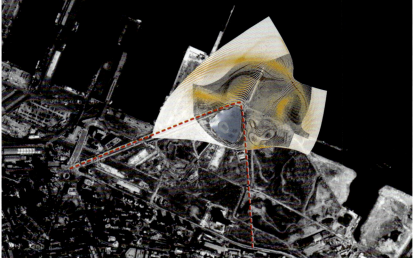

Development of the landscape

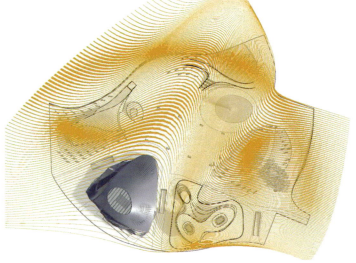
Development of the landscape

The building has both to reflect the promising modern future of Dalian and its tradition as an important port, trade, industry and tourism city. The formal language of the project combines and merges the rational structure and organization of its modern conference center typology with the floating spaces of modernist architecture.

The structural concept is based on a sandwich structure composed of 2 elements: the "table" and the roof. Both elements are steel space frames with depths ranging between 5 and 8 meters. The whole structure is elevated 7 meters above ground level and is supported by 14 vertical composite steel and concrete cores.

A doubly ruled façade structure connects the two layers of table and roof, creating a load-bearing shell structure.

The application of new design and simulation techniques, the knowledge of local shipbuilders to bend massive steel plates, and the consumption of more than 40,000 tons of steel enables breathtaking spans of over 85 meters and cantilevering of over 40 meters.

The focus of the architectural design and project development lies on technology, construction and their interplay. The technical systems fulfil the tasks required for the spatial use of the building automatically, invisibly and silently.

One of the major tasks of sustainable architecture is the minimisation of energy consumption. A fundamental contribution is to avoid considerable fluctuations in demands during the course of the day. Therefore it is essential to integrate the natural resources of the environment like:

• Use the thermal energy of seawater with heat pumps for cooling in summer and heating in winter;

• General use of low temperature systems for heating in combination with activation of the concrete core as thermal mass in order to keep the building on constant temperature;

• Natural ventilation of the huge air volumes within the building allows for minimization of the mechanical apparatus for ventilation heating and cooling. The atrium is conceived as a solar heated, naturally ventilated sub climatic area;

• In the large volume individual areas can be treated separately by additional measures such as displacement ventilation;

• A high degree of daylight use is aspired both for its positive psychological effect and for minimizing the power consumption for artificial lighting;

• Energy production with solar energy panels integrated into the shape of the building.

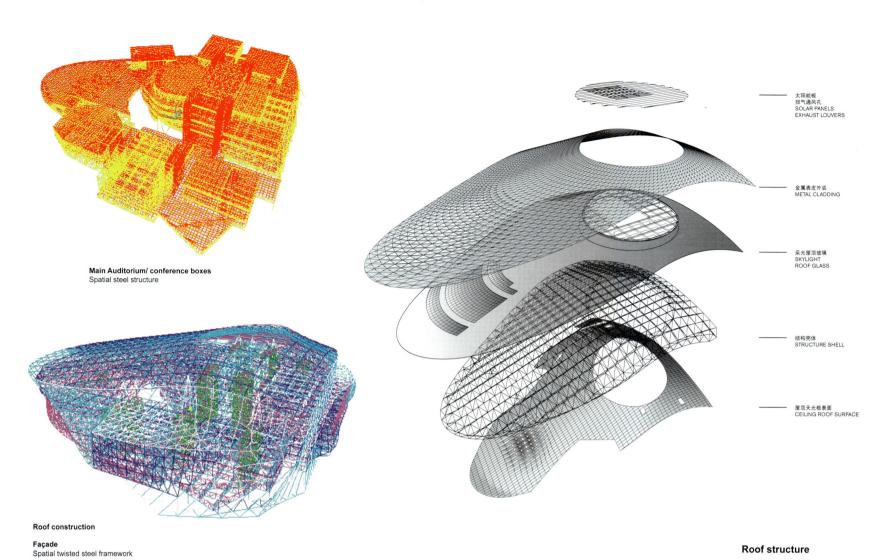

Main Auditorium/ conference boxes
Spatial steel structure

Roof construction

Façade
Spatial twisted steel framework

太阳能板
排气通风孔
SOLAR PANELS
EXHAUST LOUVERS

金属表皮外装
METAL CLADDING

采光屋顶玻璃
SKYLIGHT
ROOF GLASS

结构壳体
STRUCTURE SHELL

屋顶天光板表面
CEILING ROOF SURFACE

Roof structure

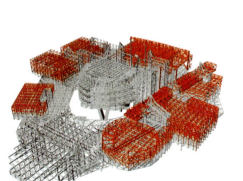

conference boxes
spatial steel structure

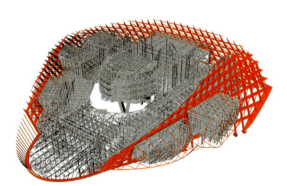

façade
spatial twisted steel framework

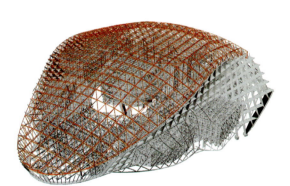

roof structure

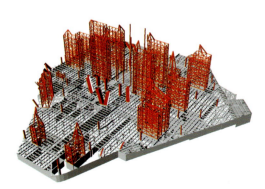

cores and columns
vertical steel concrete bond

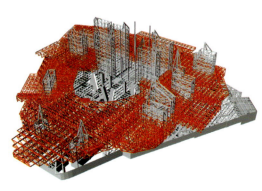

table
spatial steel framework

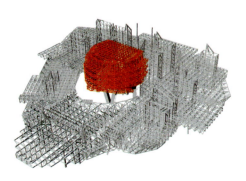

main auditorium
spatial steel structure

Program
Study model
项目研究模型

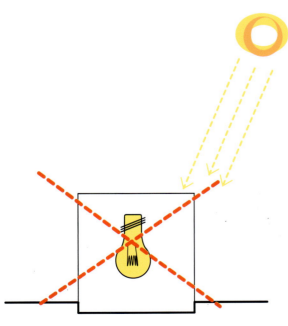

Black Box
黑盒子

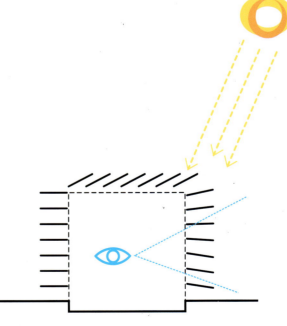

Glass Box
Façade with louvers

玻璃盒子
带有百叶窗的表皮

大连是一座集贸易、工业和旅游于一身的重要的港口城市，这座建筑既体现了大连市上述传统功能，又展现了其不可预知的发展前景。该项目的结构语言结合了现代会议中心结构的合理分布的特点，同时又体现出现代派建筑的空间流动性。

该建筑结构概念以叠层构造为基础，由"底座"和屋顶两种元素组成。两种元素均为钢化空间构架，深度在 5 米到 8 米之间。整体结构高出地面 7 米，用 14 根垂直复合钢和混凝土芯作支撑。

双层直纹立面结构连接底座和两层屋顶，打造了一个承载重量的壳体结构。

该建筑实现了超过 85 米的惊人跨度和超过 40 米的悬臂，这要归功于新设计和仿真技术的应用、当地造船专家弯曲厚钢板方面的知识，以及 4 万余吨钢材的消耗。

建筑设计和项目发展的关键在于技术、结构以及两者之间的相互影响。技术体系要在无形之中悄无声息地自动完成整栋建筑对于空间使用方面要求的任务。

可持续建筑的主要目标之一就是使能源消耗最小化。在一天之中避免较大的起伏波动是减少消耗的根本。因此，与自然环境资源融合在一起尤为重要，例如：

• 利用海水热能和热泵，夏季制冷，冬季供暖；

• 通常利用低温系统供暖，将活性混凝土芯当作蓄热体，这样便可以使建筑物保持恒温；

• 建筑物内保持自然通风，尽量减少空调的制热制冷使用。中庭则是一个由太阳热能供暖，并可自然通风的子气候区；

• 通过置换通风等额外措施，可对大量个体空间进行分别处理；

• 高度利用日照，既能产生积极的心理效应，又可以减少人工照明的能量消耗；

• 太阳能板与建筑的形状浑然一体并能产生能量。

气候设计 / CLIMATE DESIGN

海水供冷 / SEA WATER COOLING
替换通风系统 / DISPLACEMENT OF VENTILATION SYSTEMS
地板式供热及供冷 / FLOOR HEATING/ COOLING
自然通风 / NATURAL VENTILATION
太阳能光电池组 / SOLAR ENERGY/ PV CELLS
自然采光及遮阳系统 / NATURAL LIGHT/ SHADING

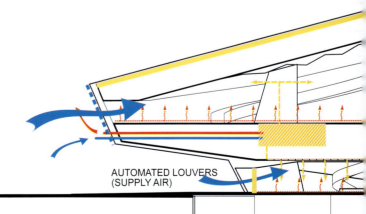

AUTOMATED LOUVERS (SUPPLY AIR)

海洋
SEA

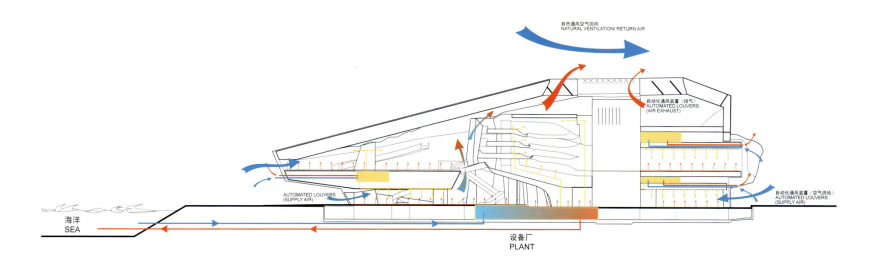

Natural ventilation
自然通风

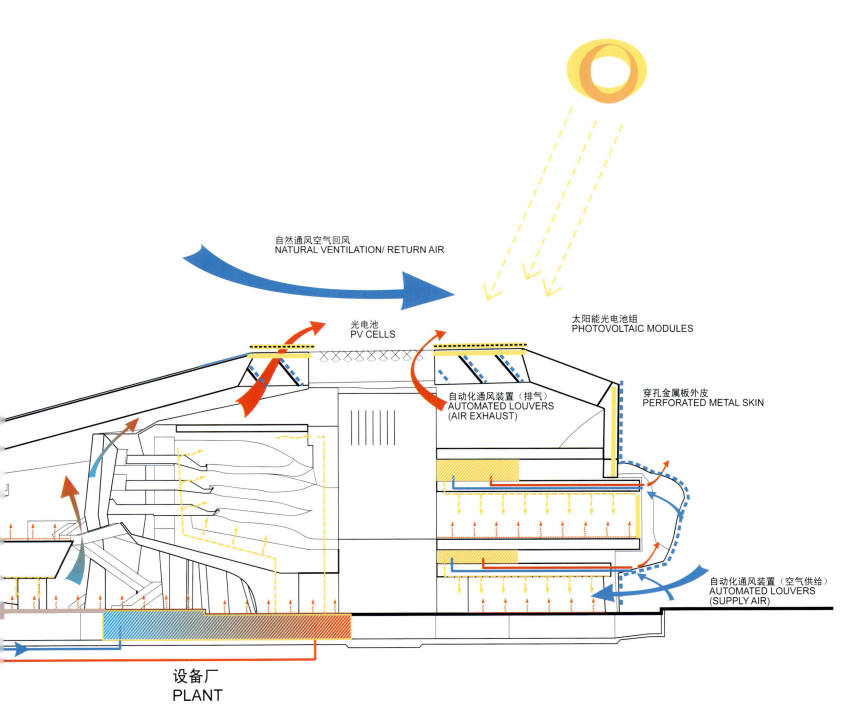

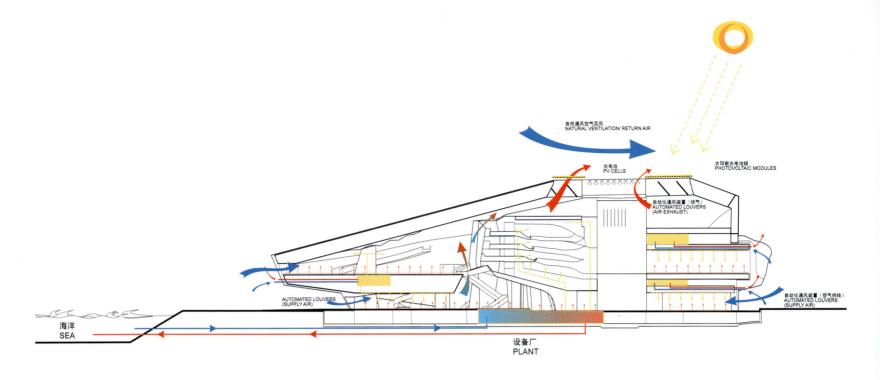

太阳能 / 光伏电池
Solar energy/ PV cells

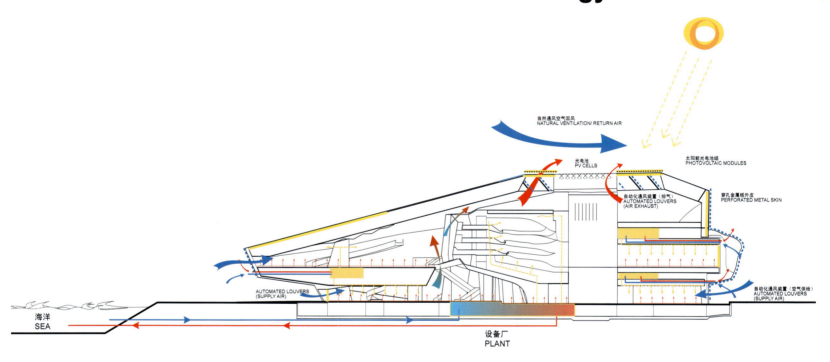

自然光 / 遮阳
Natural light/ shading

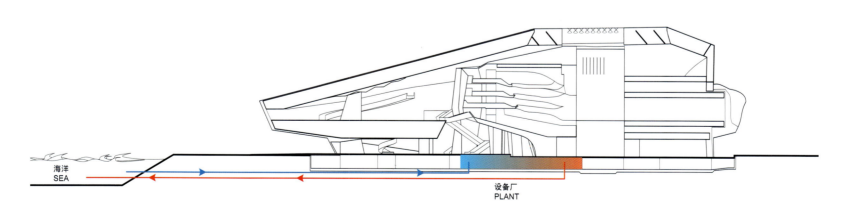

海水冷却
Sea water cooling

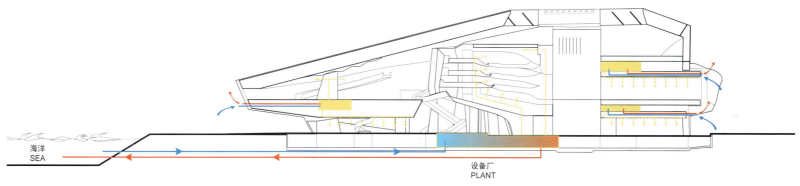

置换通风系统

Displacement of ventilation systems

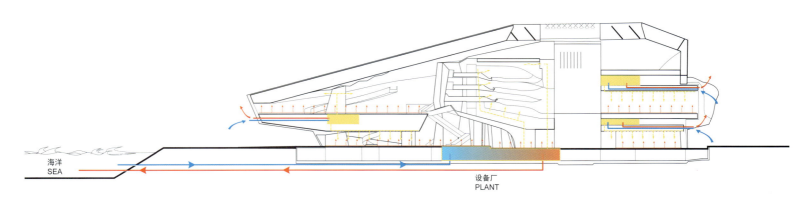

地板供暖／供冷

Floor heating/ cooling

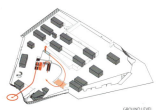

GROUND LEVEL   +15.3m LEVEL

Circulation opera

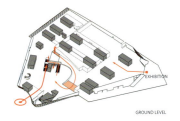
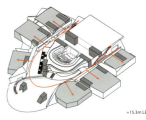
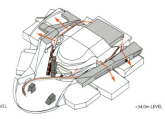

GROUND LEVEL   +15.3m LEVEL   +34.0m LEVEL

Circulation conference halls

Circulation
Study model 1:200

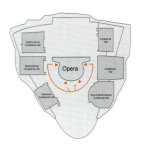

Circulation of opera and conference halls

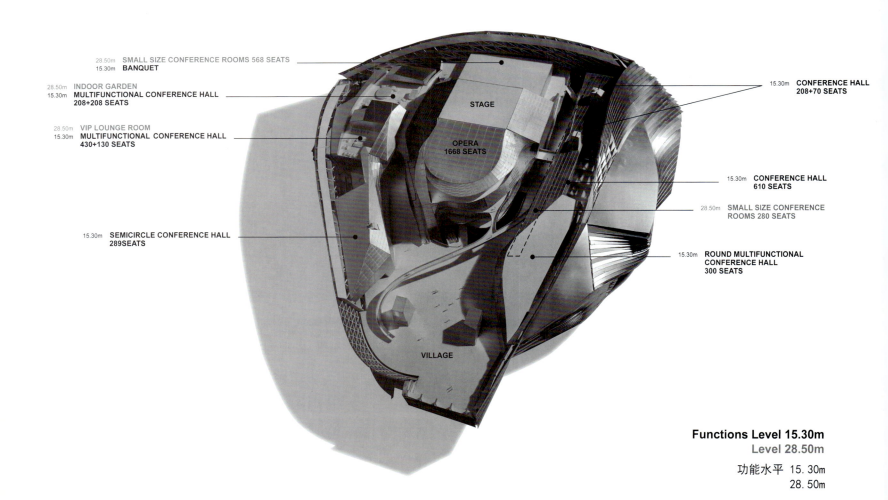

28.50m SMALL SIZE CONFERENCE ROOMS 568 SEATS
15.30m **BANQUET**

28.50m **INDOOR GARDEN**
15.30m **MULTIFUNCTIONAL CONFERENCE HALL 208+208 SEATS**

28.50m **VIP LOUNGE ROOM**
15.30m **MULTIFUNCTIONAL CONFERENCE HALL 430+130 SEATS**

15.30m **SEMICIRCLE CONFERENCE HALL 289SEATS**

STAGE

OPERA 1668 SEATS

VILLAGE

15.30m **CONFERENCE HALL 208+70 SEATS**

15.30m **CONFERENCE HALL 610 SEATS**

28.50m **SMALL SIZE CONFERENCE ROOMS 280 SEATS**

15.30m **ROUND MULTIFUNCTIONAL CONFERENCE HALL 300 SEATS**

Functions Level 15.30m
Level 28.50m

功能水平 15.30m
28.50m

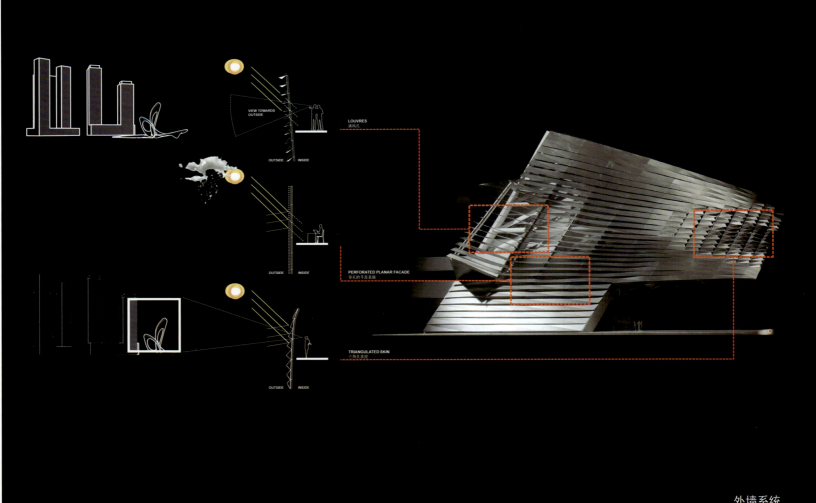

外墙系统
Façade system

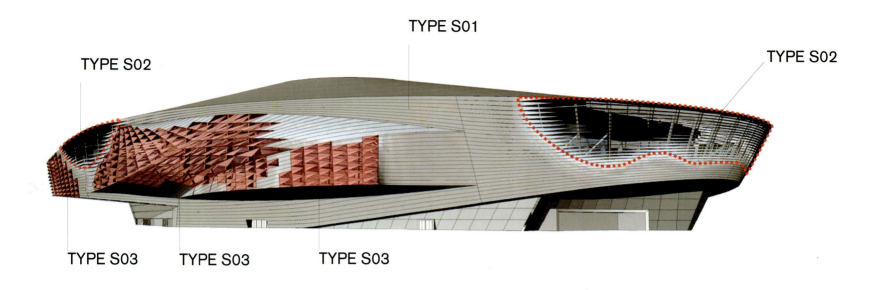

TYPE S01

TYPE S02

TYPE S02

TYPE S03

TYPE S03

TYPE S03

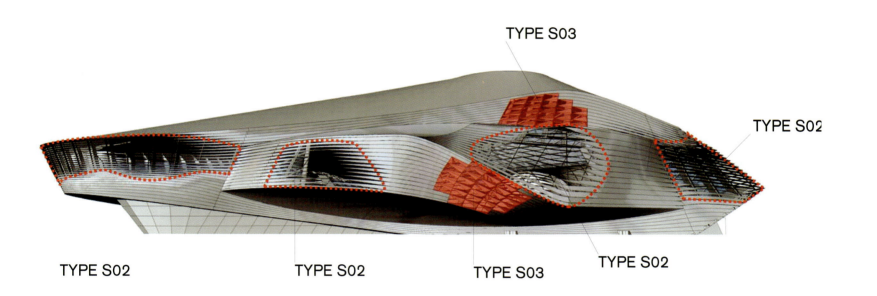

TYPE S03

TYPE S02

TYPE S02

TYPE S02

TYPE S03

TYPE S02

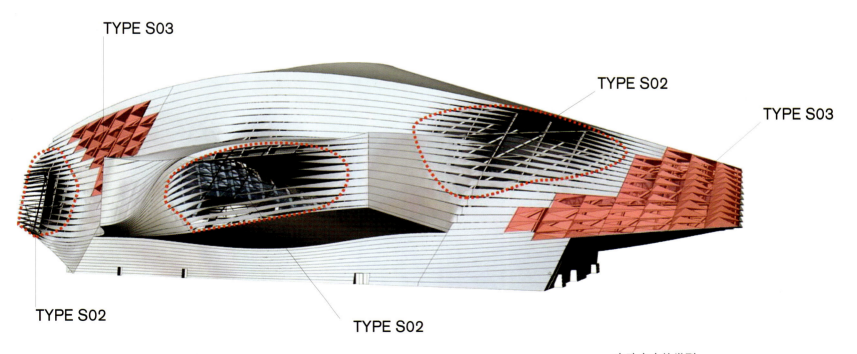

TYPE S03

TYPE S02

TYPE S03

TYPE S02

TYPE S02

穿孔表皮的类型
Types of perforated skin

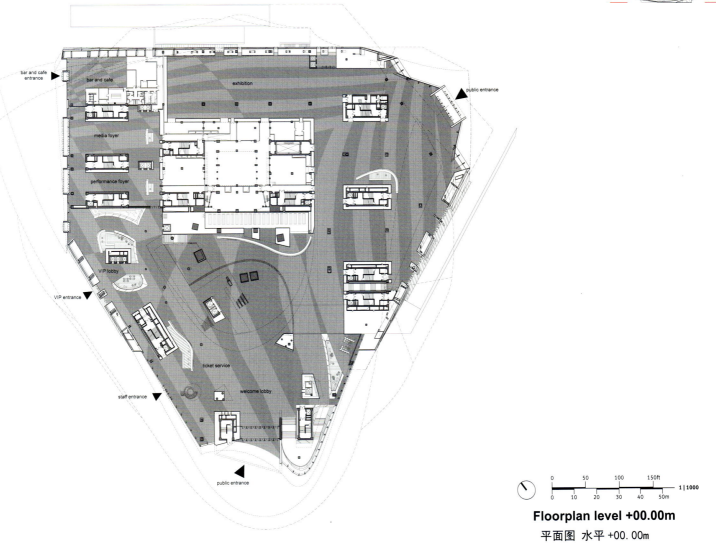

bar and café entrance

bar and café

exhibition

public entrance

media foyer

performance foyer

VIP lobby

VIP entrance

staff entrance

ticket service

welcome lobby

public entrance

0   50   100   150ft

0   10   20   30   40   50m

1 | 1000

Floorplan level +00.00m

平面图 水平 +00.00m

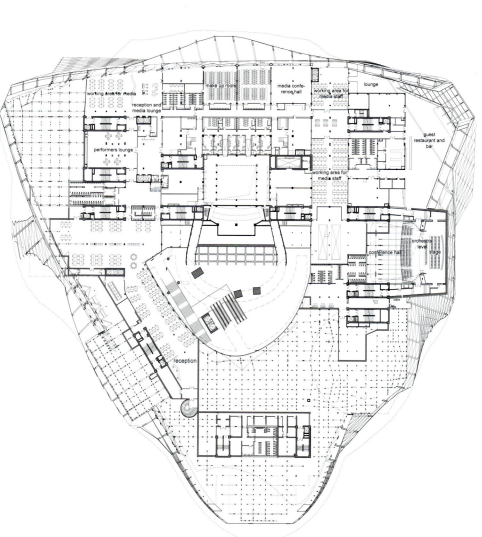

Floorplan level +10.20m
平面图 水平 +10.20m

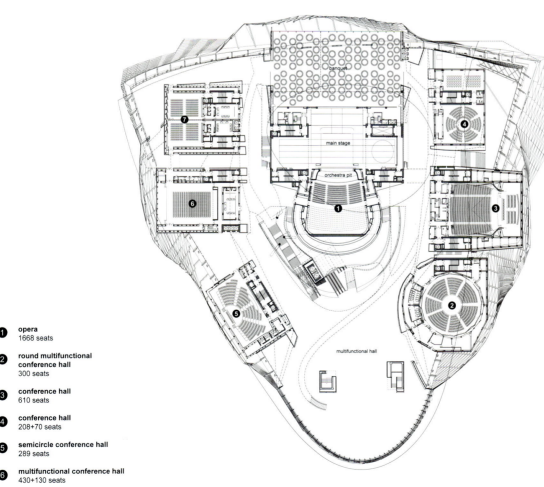

1. opera
   1668 seats

2. round multifunctional
   conference hall
   300 seats

3. conference hall
   610 seats

4. conference hall
   208+70 seats

5. semicircle conference hall
   289 seats

6. multifunctional conference hall
   430+130 seats

7. multifunctional conference hall
   208+208 seats

1 | 1000

Floorplan level +15.30m
平面图 水平 +15.30m

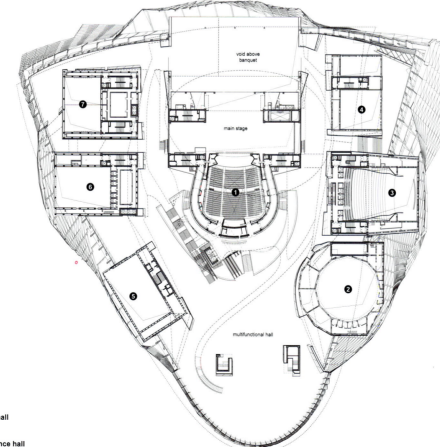

1. **opera**
   1668 seats

2. **round multifunctional conference hall**
   300 seats

3. **conference hall**
   610 seats

4. **conference hall**
   208+70 seats

5. **semicircle conference hall**
   289 seats

6. **multifunctional conference hall**
   430+130 seats

7. **multifunctional conference hall**
   208+208 seats

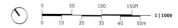

Floorplan level +17.85m
平面图 水平 +17.85m

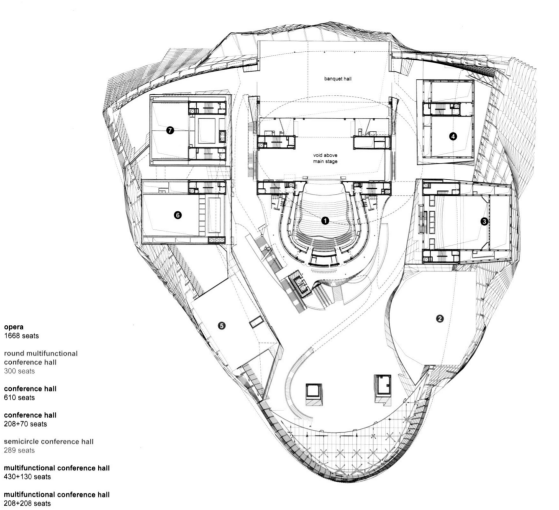

1 opera
1668 seats

2 round multifunctional
conference hall
300 seats

3 conference hall
610 seats

4 conference hall
208+70 seats

5 semicircle conference hall
289 seats

6 multifunctional conference hall
430+130 seats

7 multifunctional conference hall
208+208 seats

Floorplan level +23.00m
平面图 水平 +23.00m

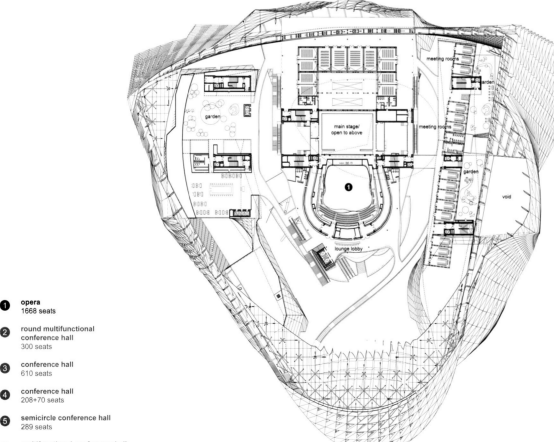

1 opera
1668 seats

2 round multifunctional
conference hall
300 seats

3 conference hall
610 seats

4 conference hall
208+70 seats

5 semicircle conference hall
289 seats

6 multifunctional conference hall
430+130 seats

7 multifunctional conference hall
208+208 seats

Floorplan level +28.50m
平面图 水平 +28.50m

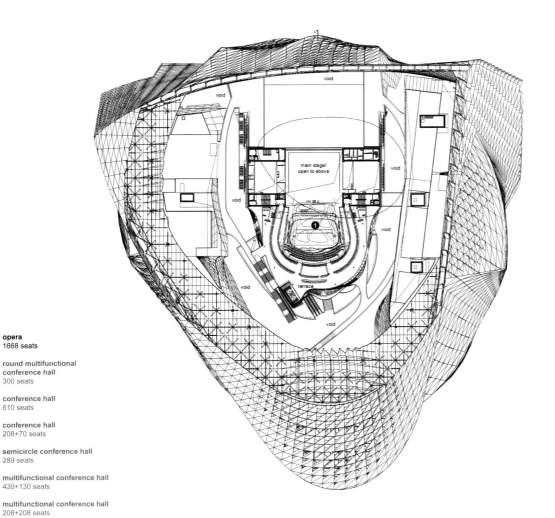

1 opera
1668 seats

2 round multifunctional
conference hall
300 seats

3 conference hall
610 seats

4 conference hall
208+70 seats

5 semicircle conference hall
289 seats

6 multifunctional conference hall
430+130 seats

7 multifunctional conference hall
208+208 seats

Floorplan level +34.00m
平面图 水平 +34.00m

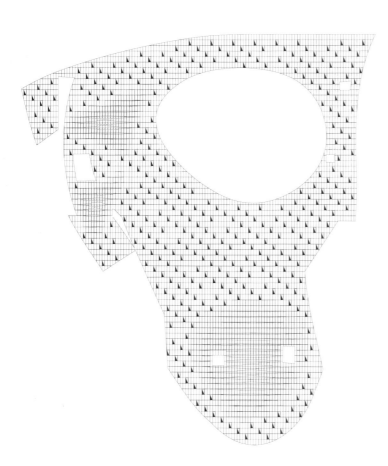

Reflected ceiling plan - Roof
屋顶天花反向图

Top view
顶视图

宴会区（1800座）
Banquet area 1800 seats

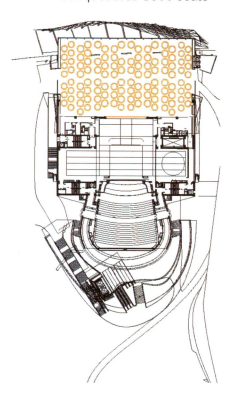

会议区（2500座）
Conference 2500 seats

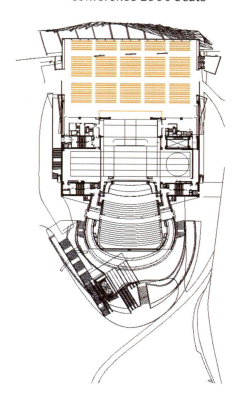

2个中型会议室
2 Midsize conference rooms

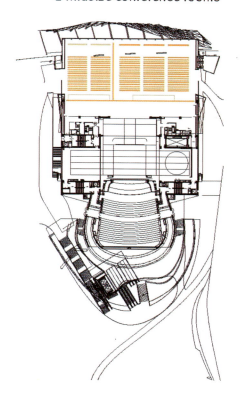

舞台中心展示区
Show-Center stage

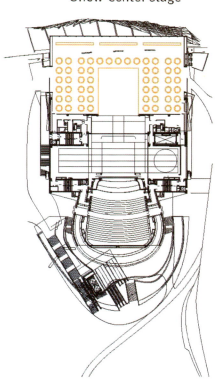

展示区
Exhibition area

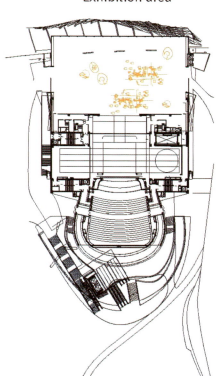

大剧院
Opera house

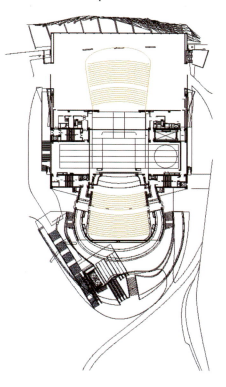

Functions Banquet
Level 15.30m
宴会厅功能分区
水平 15.30m

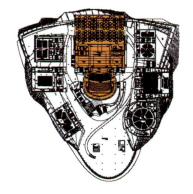

多功能会议厅（208+208 座）
Multifunctional conference hall 208 + 208 seats

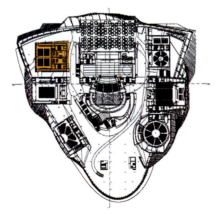

宴会厅
Banquet

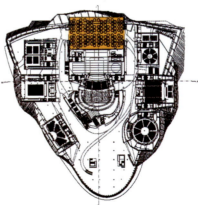

会议厅（208+70 座）
Conference hall 208 + 70 seats

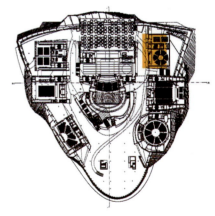

多功能会议厅（430+130 座）
Multifunctional conference hall 430 + 130 seats

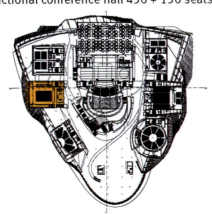

大剧院（1668 座）
Opera 1668 seats

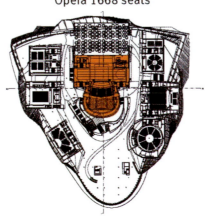

会议厅（160 座）
Conference hall 610 seats

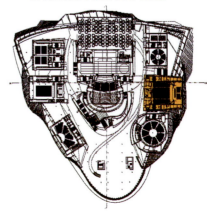

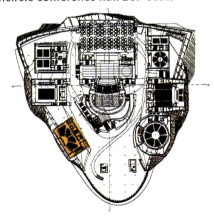

半圆形会议厅（289 座）
Semicircle conference hall 289 seats

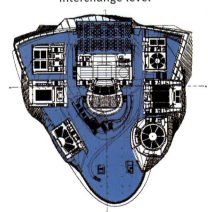

交流层
Interchange level

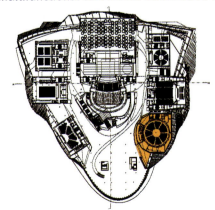

圆形多功能会议厅（300 座）
Round multifunctional conference hall mit 300 seats

会议厅、大剧院、交流层功能图
水平 15.30m
Functions
Conference halls/ Opera/ Interchange level
Level 15.30m

SHANGHAI INTERNATIONAL CRUISE TERMINAL

上海港国际客运中心

绿色特征

- *Use of seawater cooling system, breathing curtain walls and underfloor air distribution* ／ 使用江水冷却系统、呼吸式幕墙和地板送风
- *Double-layer glass façade to improve air flow and keep the building cool in summer* ／ 双层玻璃表皮改善空气流通，使大楼在夏季保持凉爽

① OFFICES PAVILIONS	③ PUBLIC GALLERY BUILDING	⑤ MEDIA GARDEN	⑦ PUBLIC PARK
② PUBLIC WINTER GARDEN	④ PUBLIC PERFORMANCE	⑥ FOOD COURT	⑧ PHASE 2 TOWERS

1. 办公展区
2. 公共冬日花园
3. 公共美术馆
4. 公共表演区
5. 媒体花园
6. 美食广场
7. 公园
8. 第 2 期塔楼

ARCHITECTURE, LANDSCAPE AND INTERIOR DESIGN
SPARK

PROJECT DIRECTOR
John Curran

TEAM
Jeb Beresford, Gabriel Briamonte, Conyee Chan, Jan Clostermann, Sofia David, Carl Harding, Zhang Hua, Ala Pratt, Joe Ren, Sven Steiner

CLIENT
Shanghai Port International Cruise Terminal Ltd

ENGINEER
Arup Hong Kong, China

FACADE ENGINEER
Arup Hong Kong, China

LIGHTING CONSULTANT
Lighting Design Partnership

LOCATION
Shanghai, China

AREA
263,448 m^2

PHOTOGRAPHER
Lin Ho, Christian Richters, Eric Chan

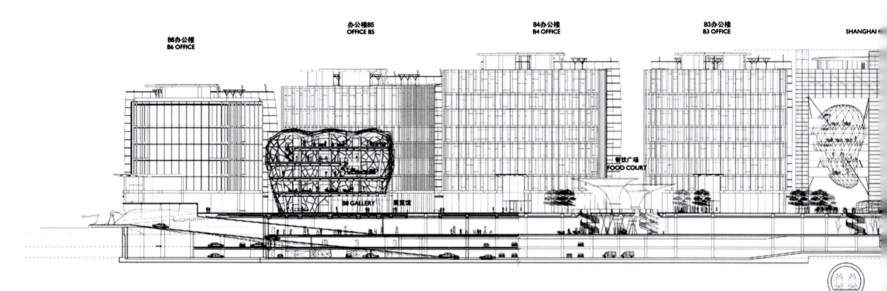

B6办公楼
B6 OFFICE

办公楼B5
OFFICE B5

B4办公楼
B4 OFFICE

B3办公楼
B3 OFFICE

SHANGHAI

一餐饮广场
FOOD COURT

B8 GALLERY 展览馆

计
PROPO

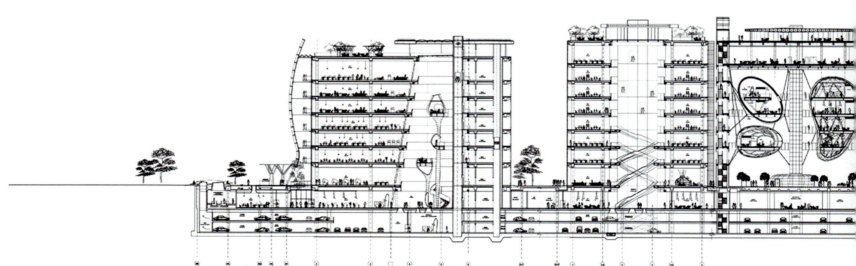

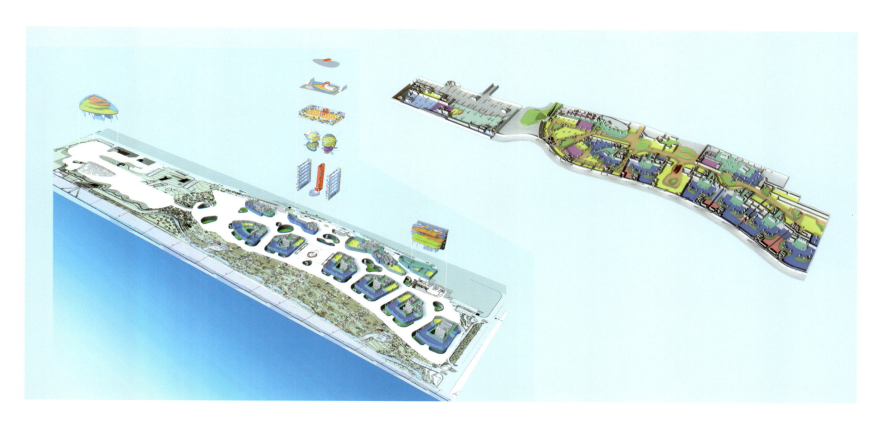

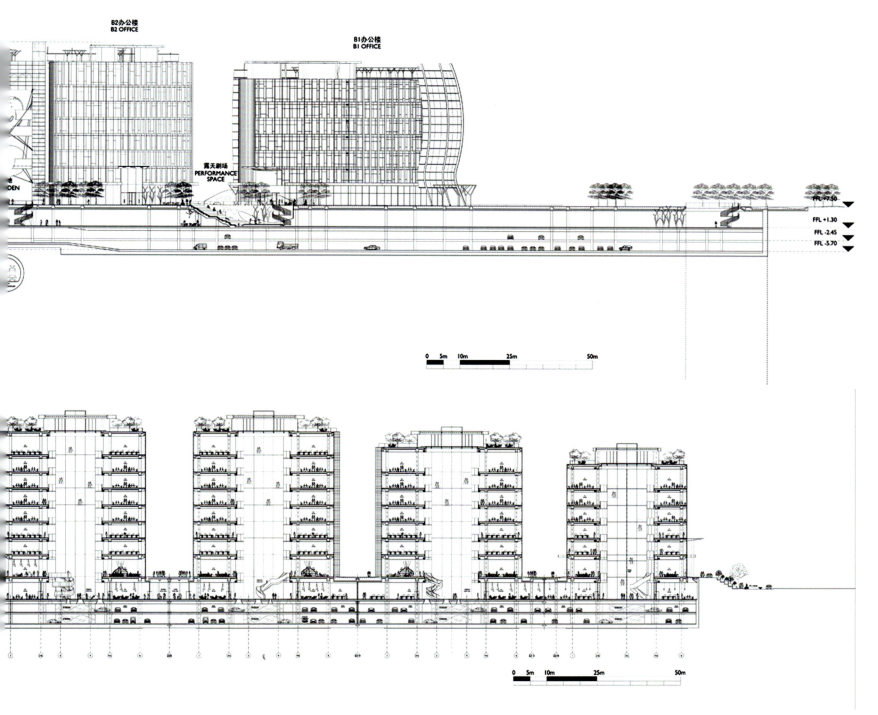

B2办公楼
B2 OFFICE

B1办公楼
B1 OFFICE

露天剧场
PERFORMANCE
SPACE

FFL +7.50

FFL +1.30

FFL -2.45

FFL -5.70

0 5m 10m   25m   50m

0 5m 10m   25m   50m

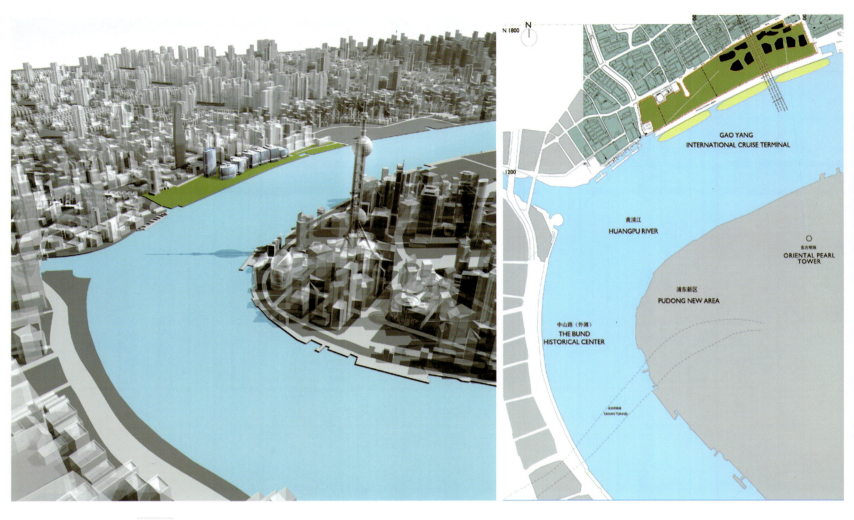

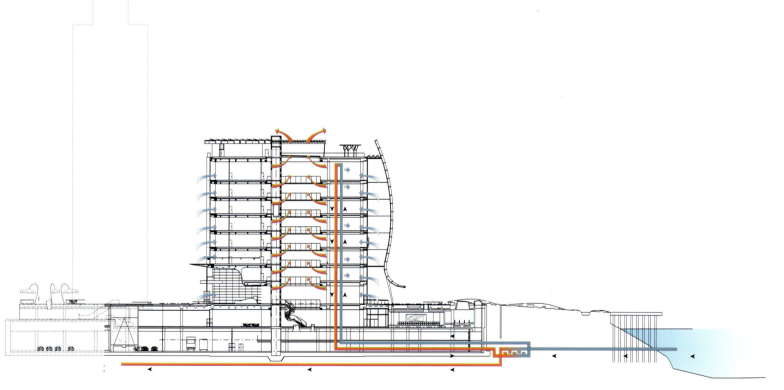

Spark was appointed the master planner of the Cruise Terminal mixed-use development in 2004. Totalling a floor area of more than 260,000 m², the development comprises 80 percent commercial use and 20 percent public facilities, and entertainment and retail outlets.

The Shanghai International Cruise Terminal offers an eclectic mix of business and entertainment to Shanghai residents and strengthened the city's status as a key commerce and tourism hub. The highlight of the development is a structure called the "Shanghai Chandelier"– a 40m-high glass-clad portal that overlooks the public park and waterfront where an open space is set aside for residents and tourists to gather for festivals and events. The structure houses several "floating" cafés, restaurants and bars suspended on cables in an extraordinary three-dimensional composition, the first suspended cable construction of its kind in the world.

The project also bears the signature imprint of Spark in marrying environmental sustainability to creative design. The development utilizes the "River Water Cooling Technology", a first in Shanghai – where water from the Huangpu River is used as a refrigerant to naturally cool the buildings during the hot summer months, and then recycled back to the river.

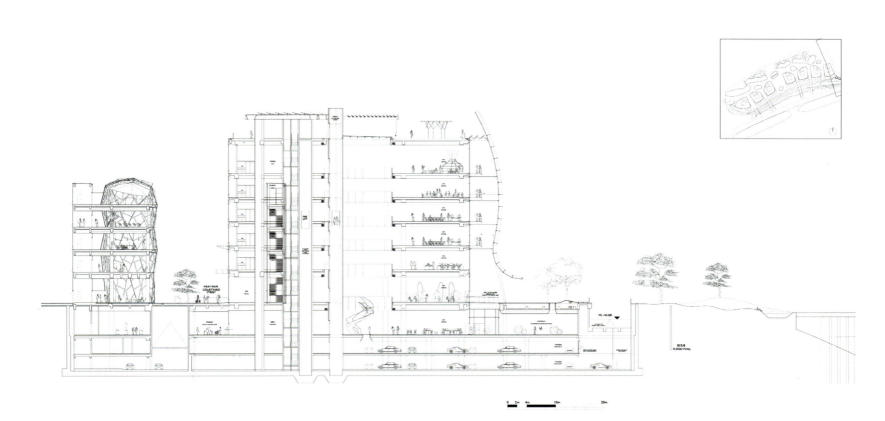

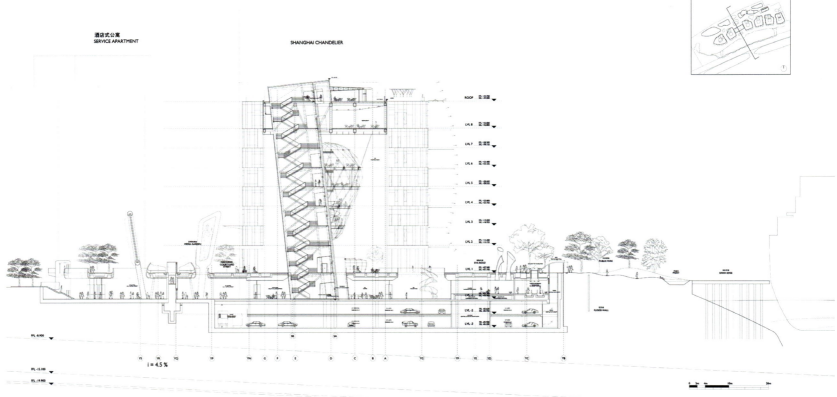

酒店式公寓
SERVICE APARTMENT

SHANGHAI CHANDELIER

2004 年，Spark 公司应邀成为上海港国际客运中心多功能开发项目的总体策划。建筑面积总计多达 26 万平方米。该项目中，商业用途占 80%，而娱乐设施、零售商店和公共设施一共占 20%。

上海港国际客运中心将商业和娱乐融为一体，不仅服务了上海居民，更增强了上海商业枢纽和旅游中心的城市形象。整个开发项目的亮点是一个名叫"上海吊灯"的玻璃入口，高达 40 米，从顶部可以俯瞰整个公园和滨水区。当遇到节日和举行盛大活动时，居民和游客可以在周边区域欢聚一堂，共同庆祝。该建筑物内部还设有"漂流"咖啡屋、餐厅和酒吧，以特殊的三围立体构造悬挂在缆绳上。它是世界上第一个使用此类结构设计的建筑。

该设计将环境持续发展性和设计创新性相结合，体现了 Spark 公司的独特理念。这是上海首个运用"河水冷却技术"的项目，以黄浦江水作为制冷剂，可以在炎热的夏天对建筑物进行天然冷却，使用之后仍可流回江中以便循环使用。

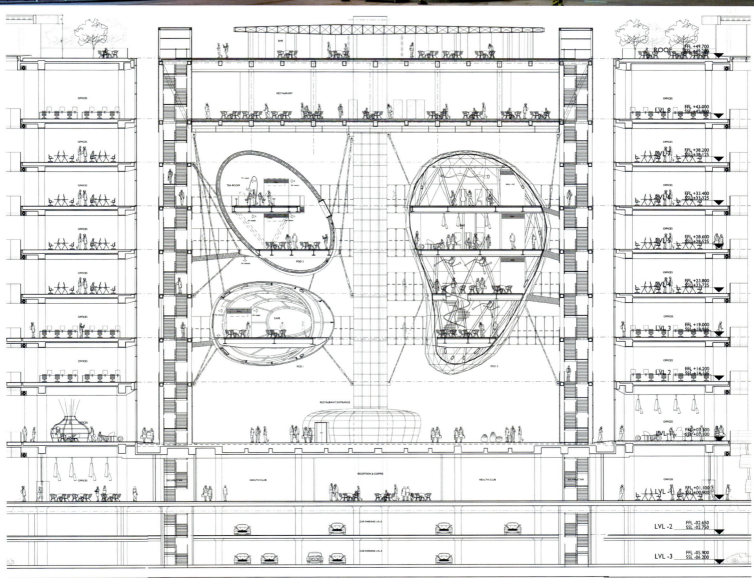

REPAS tba

零售街总平面（标高1.10米）

GENERAL ARRANGEMENT PLAN
RETAIL LEVEL 1.10M

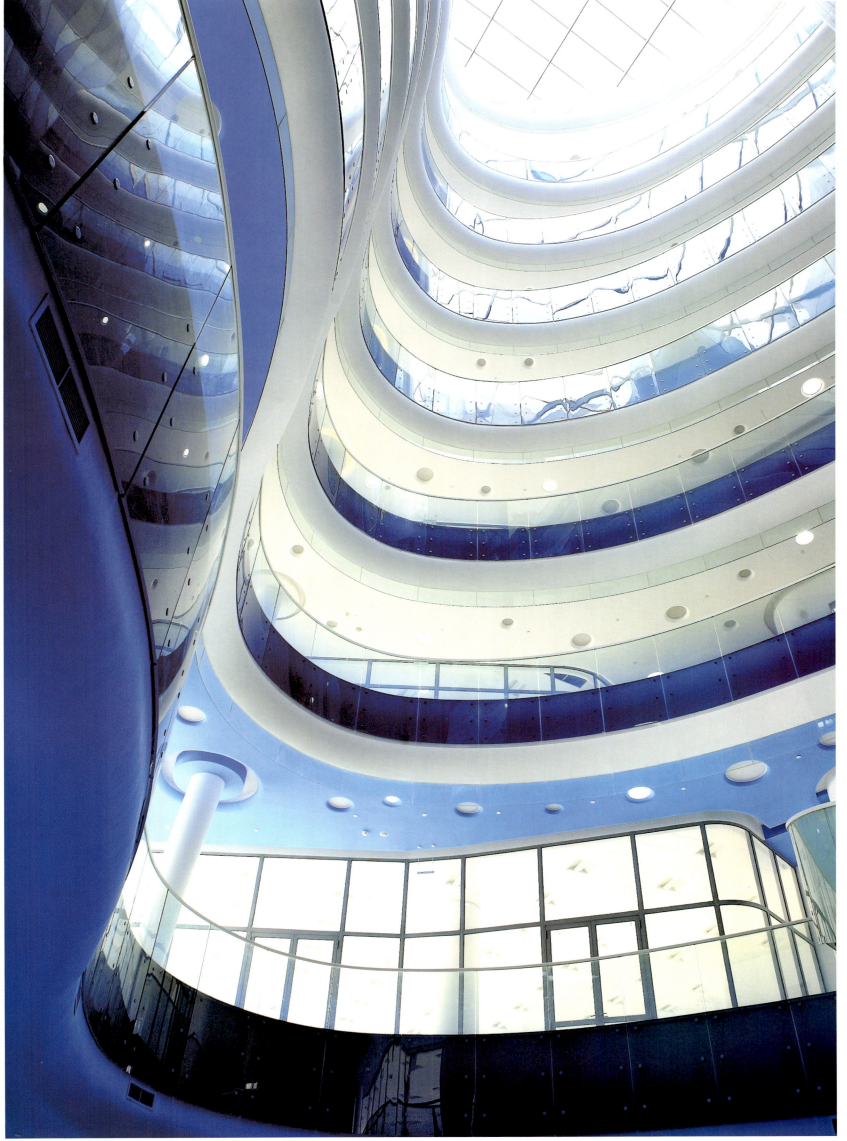

MOON BAY
INTERNATIONAL CENTER

月亮湾国际中心

CHINA GREEN BUILDING 2 STAR CERTIFIED　中国绿色建筑二星级认证

SUSTAINABLE & GREEN FEATURES　绿色特征

- *Highly efficient ventilation system* ／ 高效节能的通风系统
- *Use of double-layer and anti-radiation breathing glass curtain wall* ／ 使用双层防辐射呼吸式玻璃幕墙

ARCHITECT

Johnson Pilton Walker (Director-Jeff Walker, Associate-Carol Zhang; Project Leader: James Polyhron; Project Team: Dickson Leung, Adam Rusan, Sophie Blain, Simon Wilson, Andrew Daly, Nicholas Chou, Matthias Knauss, Elisa Nakano, Jan Wesseling, Alex Wilson, Frank Ru, Hannah Ding)

LANDSCAPE

Johnson Pilton Walker (Associate-Andrew Christie, Maggie Liang, Adam Robilliard, Adam Deutsch, Josh Harold, Maria Rigoli), Hezhan

CIVIL, COMMUNICATIONS, ELECTRICAL, MECHANICAL, STRUCTURAL, AND HYDRAULIC CONSULTANTS, CONSTRUCTION DOCUMENTATION

Suzhou Industrial Park Design and Research Institute, Key Personnel: Shi Ming, Li Genmin, Li Zheng, Zhang Yun, Zhang Kai, Mou Deya

CURTAIN WALL CONTRACTOR

Shenzhen Jin Yue Curtain Wall Company

LOCATION

Suzhou, Jiangsu Province, China

AREA

80,000 m²

PHOTOGRAPHER

Yao Li

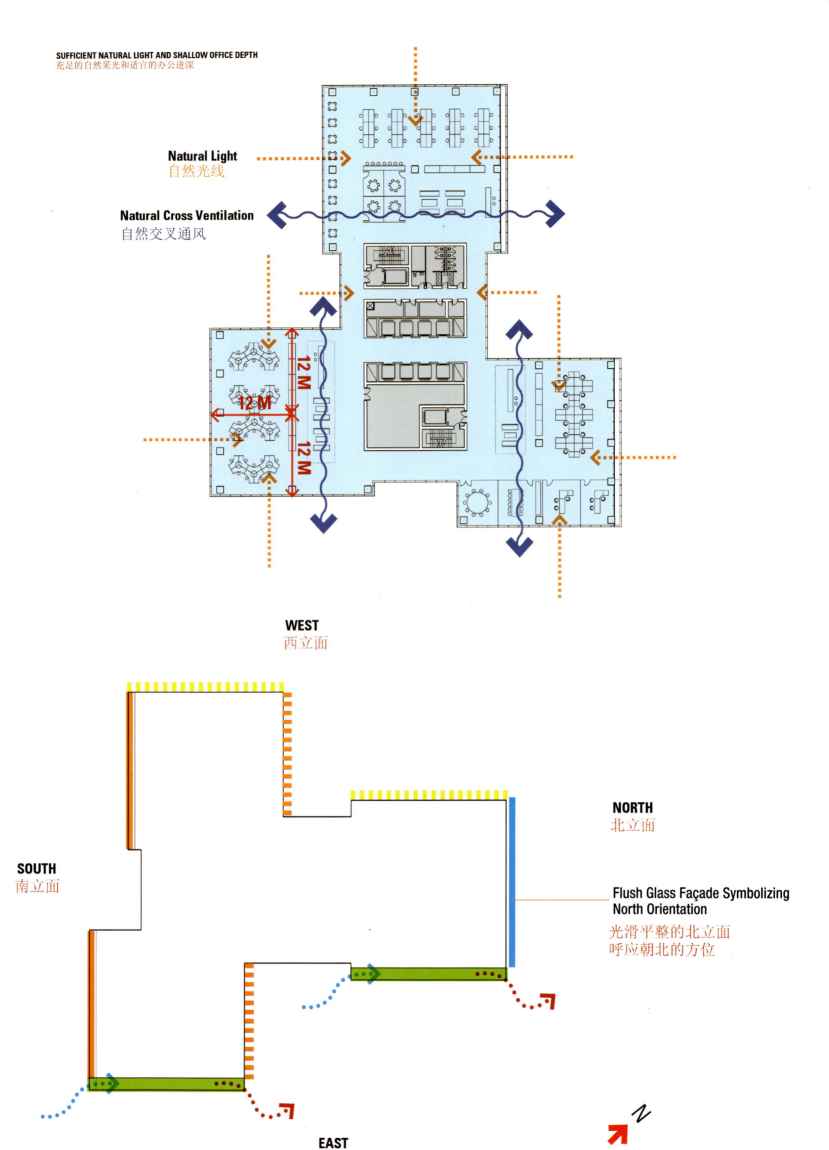

SUFFICIENT NATURAL LIGHT AND SHALLOW OFFICE DEPTH
充足的自然采光和适宜的办公进深

Natural Light
自然光线

Natural Cross Ventilation
自然交叉通风

12 M
12 M
12 M

WEST
西立面

SOUTH
南立面

NORTH
北立面

Flush Glass Façade Symbolizing North Orientation

光滑平整的北立面
呼应朝北的方位

EAST
东立面

N

As the northern gateway to the Moon Bay precinct of Suzhou Industrial Park, Site 337 is a key site that establishes the identity and character of Moon Bay's unique waterfront location. 360 degree views extend from the lake to the west, parkland to the north, the education precinct to the east and future Moon Bay development sites to the south.

The landscape design integrates the river and site pedestrian networks, encouraging pedestrians to enter the site from all directions and enjoy the green landscaped environment and variety of commercial facilities.

The building form, with a distinct podium and tower, allows each element to create its own strong identity, within an overall harmonious composition.

The Podium building encapsulates a series of landscaped terraces stepping down from east to west, extending internal functions as indoor/outdoor recreation and entertainment spaces overlooking the canal and parkland beyond.

Each tower façade has a unique characteristic responding to its solar orientation, with each unified by shared details, materials and colors. Sunhoods on the south-west façades and aluminium inlays on the north-east façades of the two tallest towers form a pattern reflecting an imagery of sails. Lighting is integrated into these horizontal elements forming a distinctive pattern at night.

Dual walls on the south-east façades incorporate vertal and horizontal glass and aluminium fins providing sun protection to the interior office areas. The pattern created by this framework relates to traditional Suzhou garden wall screens.

Mirrored glass panels provide play of reflected light during the daytime, an effect similar to sparkling water. The façade pattern is further enhanced by the introduction of colored glass panels which, when illuminated at night, present a random pattern of colored light-like reflections from the rippled water on the lake surface.

Whilst the office floors offer large floor plates, these are divisible into smaller modules, if required, and all spaces enjoy high quality natural light. This strategic approach to floor space planning and a range of other sustainable design principles have enabled Site 337 to achieve 2 star design certification.

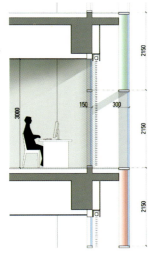

EAST FACADE DUAL WALL SECTION
东立面双墙系统剖面

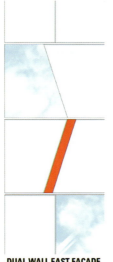

DUAL WALL EAST FACADE
东立面-双墙系统

WEST FACADE WALL SECTION
西立面幕墙系统剖面

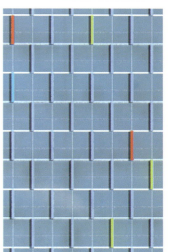

SUNHOOD ON WEST FACADE
西立面的遮阳竖挺

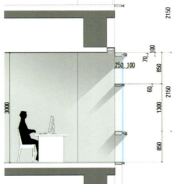

NORTH FACADE WALL SECTION
北立面幕墙剖面

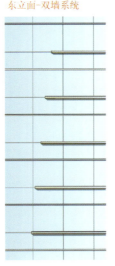

SOUTH FACADE HORIZONTAL SHADING BLADES
南立面水平遮阳板剖面

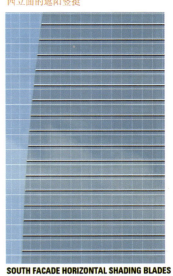

SOUTH FACADE HORIZONTAL SHADING BLADES
南立面水平遮阳板剖面

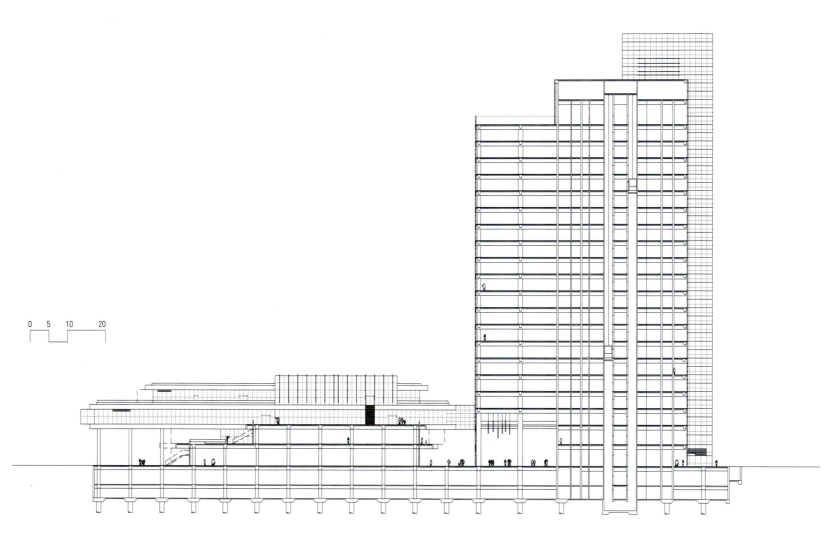

0 5 10   20

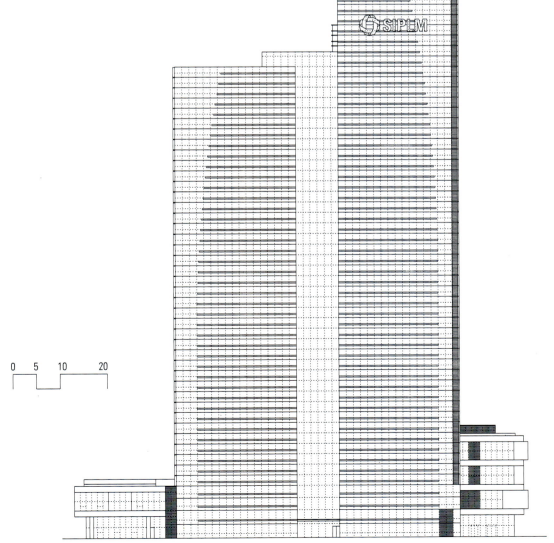

337 地块是通往月亮湾的北门户，是建立月亮湾独特滨水地理位置和滨水特性的重要地块。此地块西临湖边，北临公园，东临教育园区，南临未来月亮湾发展区，呈现 360° 全景美景。

景观设计将河流与人行步道网络结合在一起，鼓励行人从四面八方走到这里，尽享绿色景观环境及各种商业设施。

该建筑由独特的裙房与塔楼构成，地块中的每个建筑形态都具有自身强烈的可识别性，但整体上又和谐统一。

裙房自东向西设有退台式绿化露台，可作为室内／外娱乐消遣空间，拓展了室内功能，也可从此处俯瞰远处的运河与风景区。

每栋塔楼的外观立面都按照朝向不同而别具一格，但各塔楼在细节、材料及颜色方面却是相同的。在两栋最高的塔楼中，西南立面上的外遮阳板与东北立面上的铝制镶嵌形成了一个类似帆船的图案。在夜间，灯光结合水平遮阳板，形成了与众不同的图案。

东南立面的双层幕墙系统采用竖向与横向的玻璃与铝制鳍片，为室内办公区域遮阳。这一结构借鉴了传统的苏州园林幕墙。

在白天，镜子般的玻璃嵌板具有反射阳光的作用，如湖面耀动的水纹一般。同时，彩色玻璃嵌板的运用也使立面的图案更加完美。夜晚在灯光的照射下，嵌板就会随机呈现出霓虹灯般的图案——宛如湖面上粼粼水波的倒影。

办公区平面设计是把一个较大的办公平面按需要分隔成更小的单元模块。所有空间均可享受高质量的自然光。精良的建筑规划设计和一系列可持续设计措施使 337 地块获得了两星级设计认证。

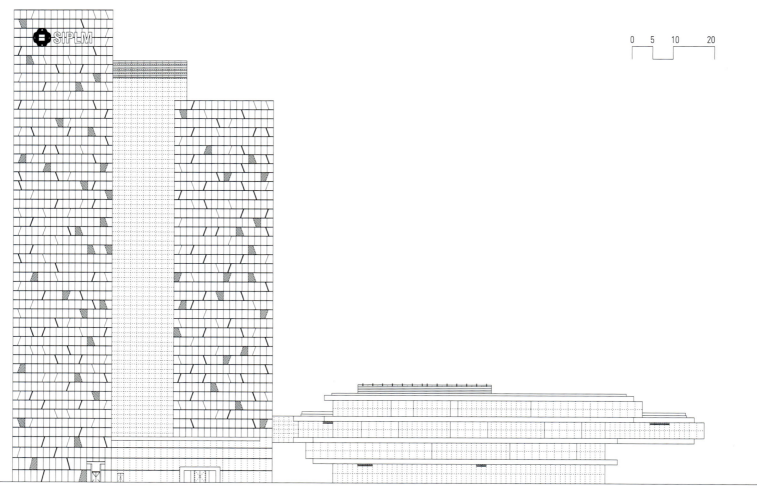

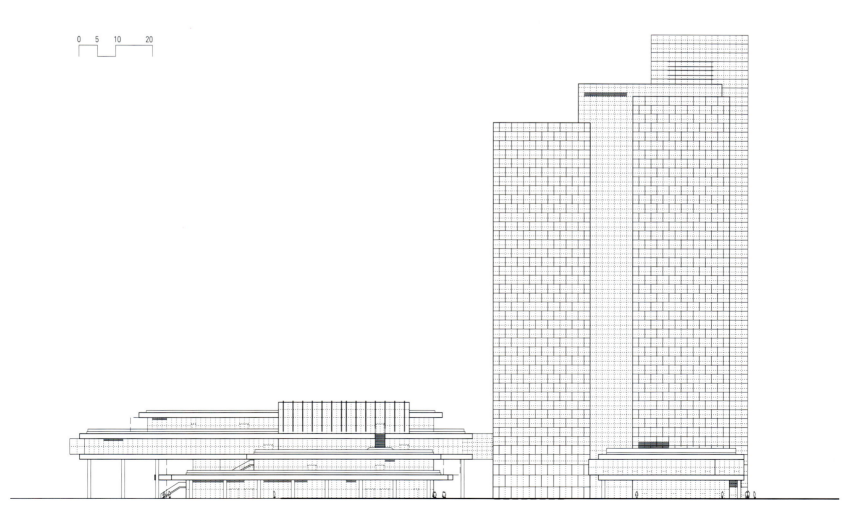

0   5   10   20

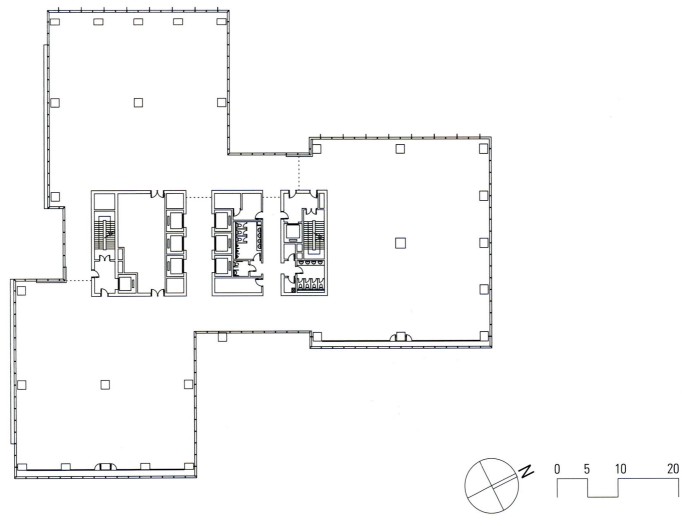

HENDERSON METROPOLITAN 恒基名人购物中心

BEAM SILVER AND LEED CERTIFIED 香港绿色建筑银级认证和美国绿色建筑认证

ARCHITECT
Tange Associates

LOCATION
Shanghai, China

AREA
1,001.60 m² (Site), 5,026 m² (Building), 67,789 m² (Total Floor)

PHOTOGRAPHER
T+E Image Architects

SUSTAINABLE & GREEN FEATURES 绿色特征

• *Low-E glass curtain wall* ／ 使用中空 Low-E 玻璃幕墙

• *Household metering central air-conditioning system* ／ 分户计量的中央空调系统

• *High isolation and noise insulation* ／ 高度隔热、隔音

The design of the Henderson Metropolitan Building is inspired by the flow of Shanghai's Huangpu River, one of the city's most important features. The building is located near the water and is clearly visible from across the river from the Pudong side of the city. On an urban scale, the undulation of the façade follows the movement and flow of the water and emphasizes the connection between the building and the Huangpu. The sunlight reflects on the glass and aluminum façade, creating a sense of dynamic movement as the light changes through the course of the day.

Henderson Metropolitan is an environmentally conscious mixed-used building, designed in consideration of both the internal environment and the surrounding environment. Our challenge was to ensure that the building would establish itself as a landmark, but given the height restriction of only 100m and the need to maximize the volume, the building risked becoming box-like. We successfully accomplished this mission by creating a dynamic façade, enabling the building to convey a stylish image. This area is a pedestrian shopping area, where the external façades of the majority of the buildings are covered with commercial ads. There is no prominence, no uniqueness. The buildings express and distinguish themselves only by their ads. The Henderson Metropolitan building also contains ads, but they are well integrated in the architecture and this integration introduces a new typology and new solution of the retail/office/mixed-use building.

The façade is an aluminum curtain wall characterized by vertical fins of varied depth between 300 and 800mm. The fins provide a visual variation, while at the same time, improving the work environment inside by softening the late afternoon sun and the exterior environment by cutting down the reflected light. Additionally, to maximize the character of the site and to avoid obstructing the view of the Huangpu River, a water-drop shaped void is applied to the deep part of fin.

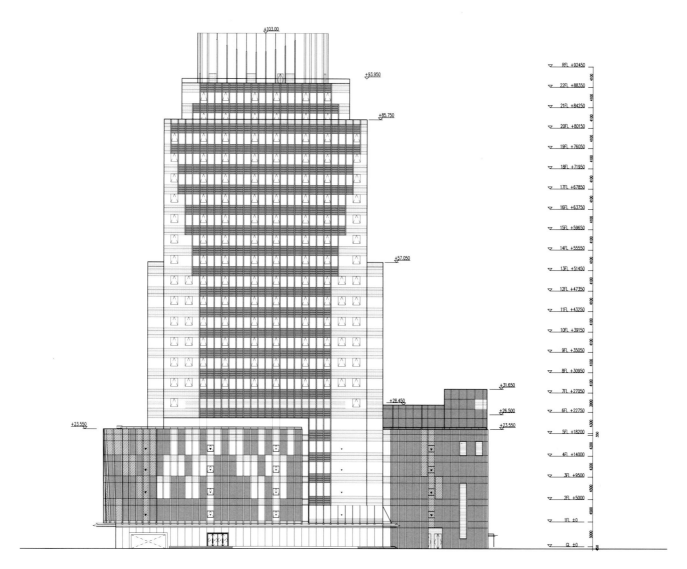

North Elevation　北外立面图　1:500（A3）

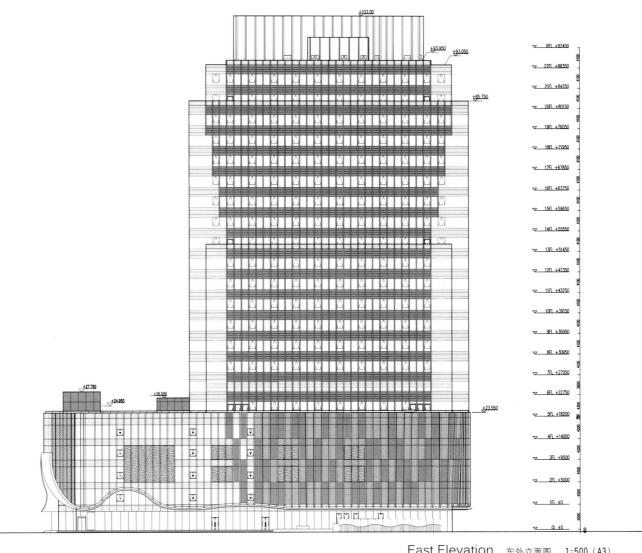

East Elevation　东外立面图　1:500（A3）

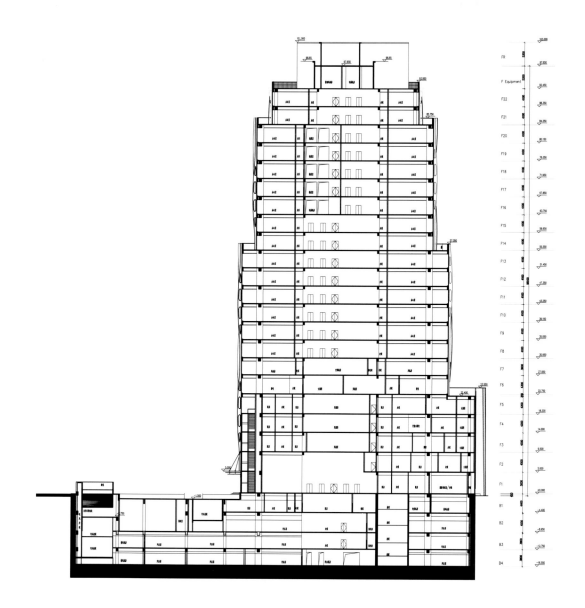

Section
剖面图　1:600（A3）

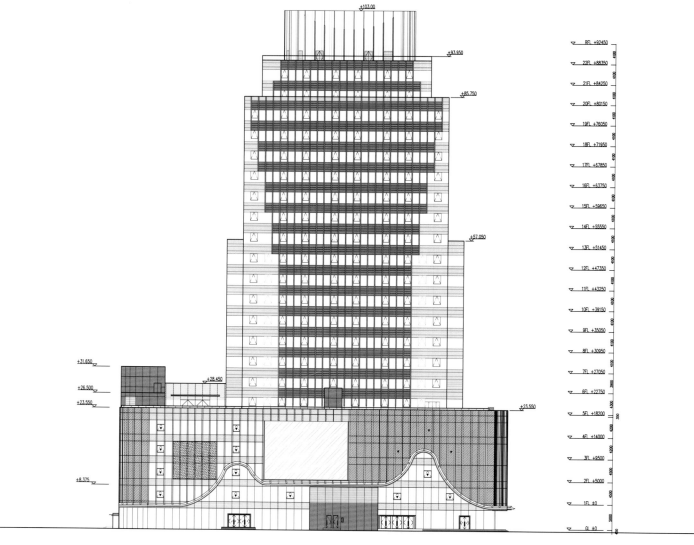

South Elevation　南外立面图　1:500（A3）

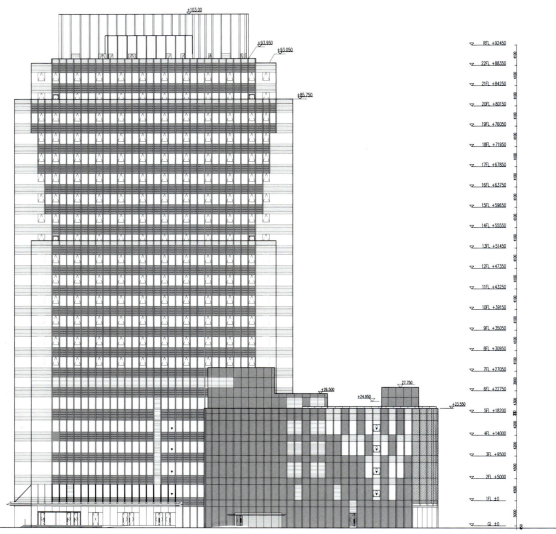

West Elevation 西外立面图 1:500 (A3)

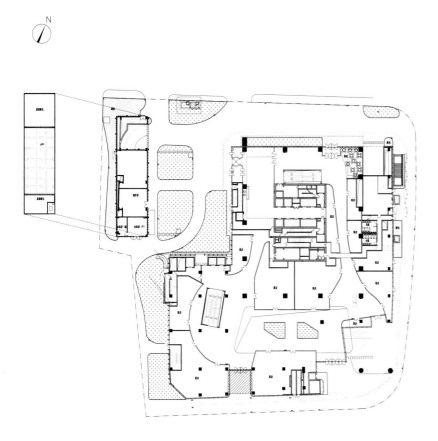

Ground Floor Plan   底层平面图   1:500（A3）

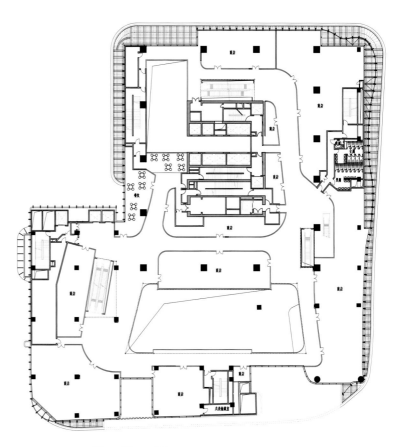

First Floor Plan   二层平面图   1:500（A3）

黄浦江是上海非常重要的特征之一，恒基名人购物中心的设计灵感正是来源于这条川流不息的河流。该购物中心坐落在河畔，可以清楚地看到河对岸的浦东区。建筑外观呈起伏状，模仿了波动流淌的河水，强调了购物中心与黄浦江的联系。阳光映射在玻璃和铝合金墙面上，营造出灵动之感，随着一天中光线的变化而产生不同的效果。

恒基名人购物中心是一栋多功能建筑，环保意识极强，设计中不仅考虑到室内环境，而且还将周遭环境也考虑在内。我们面临的挑战是，在有100米高度限制和空间需求最大化的情况下，冒着大楼会变成盒状建筑的风险，将它建成一座地标式建筑。通过建造动态立面，赋予建筑时髦的形象，我们最终成功地完成了这一任务。

该建筑所在地点是步行街，大部分建筑的外墙上都挂着商业广告。这样一来，建筑就既不显眼，又不独特，只能通过它们的广告来显示各自的不同。恒基名人购物中心也有广告，但是这些广告同建筑融合在一起，浑然一体。这种一体化的方式赋予了集零售、办公等多功能于一身的建筑以新类型和新方案。

建筑外立面采用铝制幕墙，镶有300至800毫米深度不等的垂直安定面。这些垂尾翼形成不同的视觉效果，同时使午后稍晚时候的阳光变得柔和，改善了室内的工作环境，而且也能通过减少反射光使外部环境得到改善。除此之外，为了最大程度地突出建筑地址的特色，使人们可以纵览黄浦江的美景，设计还在安定面较深的地方构筑了水滴形空间。

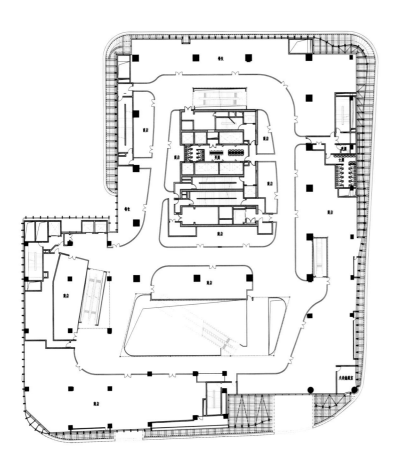

Second Floor Plan   三层平面图   1:500（A3）

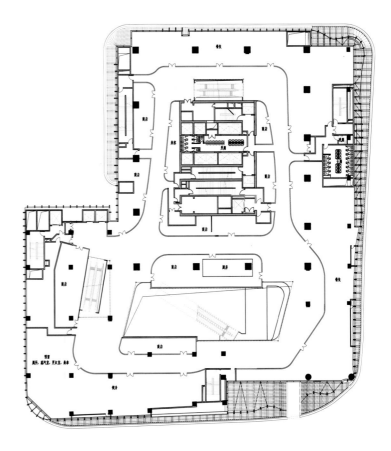

Third Floor Plan   四层平面图   1:500（A3）

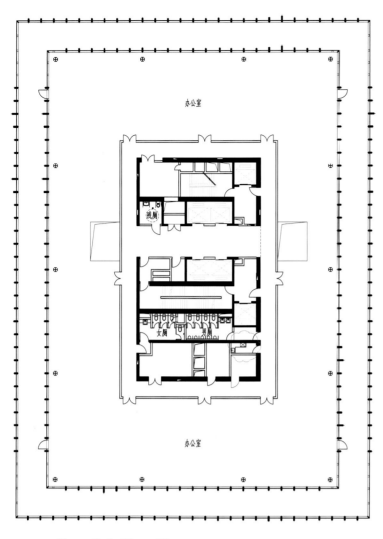

Twentieth Floor Plan　二十一层平面图　1:300（A3）

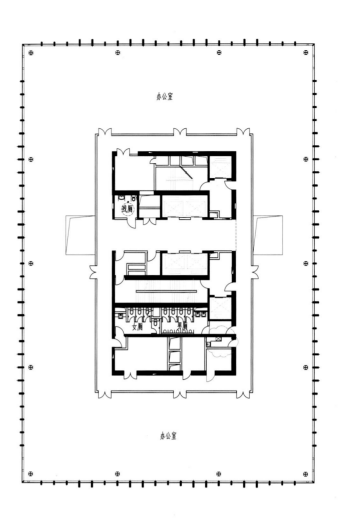

Twenty-First Floor Plan　二十二层平面图　1:300（A3）

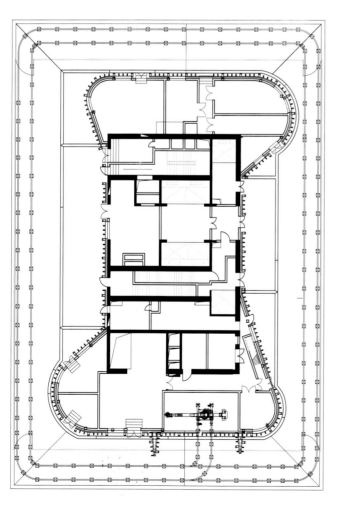

Machine Room Plan　机房层平面图　1:300（A3）

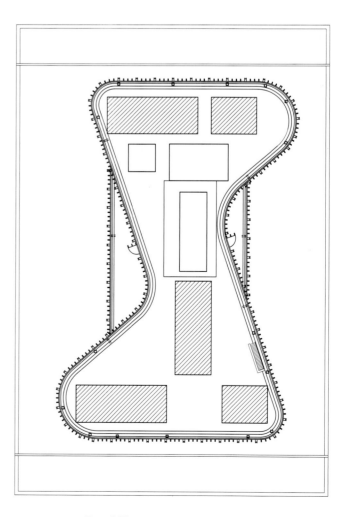

Roof Plan　屋顶层平面图　1:300（A3）

TAIKOO HUI

太古汇

LEED GOLD CERTIFIED　美国绿色建筑金奖认证

SUSTAINABLE & GREEN FEATURES　绿色特征

- *Operable indirect lighting system to reduce energy consumption* ／ 可调节的间接照明系统减少能源消耗
- *Large double-layer Low-E glass curtain wall to ensure daylight, heat preservation, heat insulation and noise insulation, and to reduce energy consumption* ／ 大型双层低辐射玻璃幕墙确保自然光穿透，同时减少能源消耗，并有助保温、隔热以及隔音
- *Energy-saving grey water recycling system* ／ 节能的中水回收系统
- *Green roof to recycle rainwater and lighten heat island effect* ／ 绿色屋顶用以雨水回收，减轻热岛效应

ARCHITECT
Arquitectonica

PARTNERS-IN-CHARGE OF DESIGN
Bernardo Fort Brescia, Laurinda Spear

ARCHITECT OF RECORD
Guangzhou Design Institute

ASSOCIATE ARCHITECTS
LWK Architects Ltd

INTERIOR DESIGNER
Arquitectonica (Retail), Tony Chi
(Mandarin Oriental Hotel)

CLIENT/OWNER
Swire Properties Limited, Taikoo Hui (Guangzhou) Development Company Limited

LOCATION
Guangzhou, Guangdong Province, China

AREA
457,000 m²

An about 450,000 m^2 mixed-use project developed by Swire Properties, the development includes 160,000 m^2 of office; 120,000 m^2 of retail; 65,000 m^2 of hotel with a 5-star, 286-key Mandarin Oriental Hotel and serviced apartments; and a 60,000 m^2 cultural center with a 1,000-seat performing arts center.

TaiKoo Hui office building won the pre-certification of LEED Gold Certificate awarded by the American Association of Green Building. This is the first building in Guangzhou to have received the honor, which affirms the green building achievements of TaiKoo Real Estate and TaiKoo Hui. TaiKoo Hui will encourage and support the office tenants to participate as candidates for LEED CI.

• Adjustable indirect lighting system, reducing up to 60% of the interior dazzle and energy consumption;

• Extra large double Low-E glass curtain wall that insures natural light and at the same time reduces energy consumption and helps with sound and thermal insulation;

• Utilizing the advanced catalytic technology of optical/NM photonic purification system to improve the interior air quality;

• Water recycling system that turns the recycled water into toilet water;

• CO_2 sensor that can moderate fresh air supply, making sure that the air is pure and fresh meanwhile avoiding the over consumption of energy;

The landscaped park also acts as a green roof insulating the retail

podium below, a sustainable landscape design solution to the requirements of a public park. The podium park serves several purposes in the landscape design: Rainwater retention, providing building insulation, reduction of energy usage, improved air quality, creating a habitat for wildlife, helping to lower urban air temperatures, and combating the heat island effect.

These sensitive design responses in turn improve on energy performance of the development. When this is promoted together with the use of local indigenous trees and vegetation, recycled content materials and construction waste management, it promotes good sustainable landscape design practice. The principles of which are shown in an interplay of the landscape forms within the architectural and urban context.

EAST ELEVATION
东立面

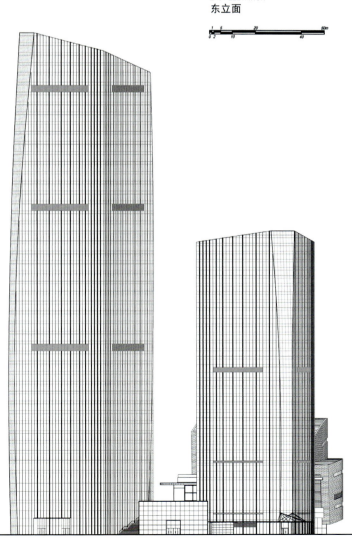

北立面
NORTH ELEVATION

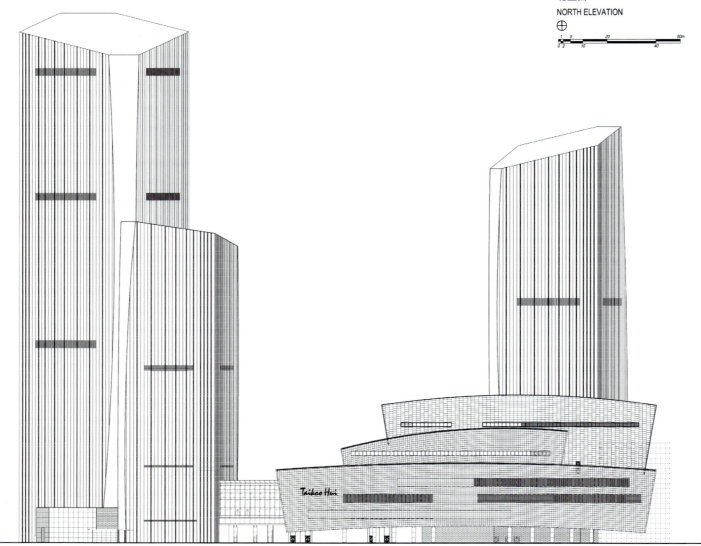

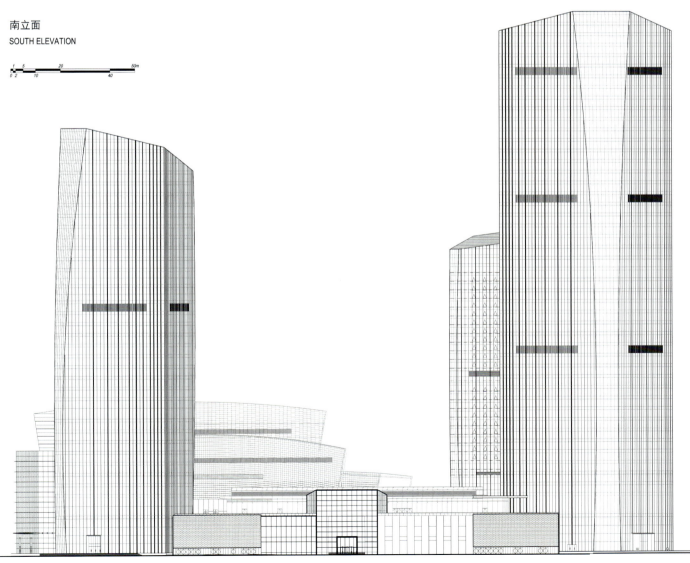

南立面
SOUTH ELEVATION

WEST ELEVATION

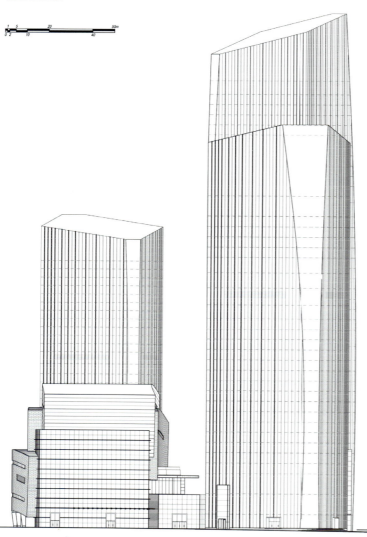

该多功能设计工程由太古地产开发，占地大约 45 万平方米。项目包括占地 16 万平方米的办公楼；占地 12 万平方米的商场；占地 6.5 万平方米的五星级 286 室的文华东方酒店和酒店式公寓；占地 6 万平方米的文化中心和能坐下 1000 人的表演文化中心。

太古汇写字楼获得了美国环保建筑协会颁发的 LEED 环保评级金奖前期认证。这是广州第一座获得 LEED 环保评级金奖的写字楼，进一步肯定了太古地产以及太古汇的绿色建筑成效。太古汇亦将鼓励和支持写字楼租户申报 LEED 室内设计类奖项评级。

• 可调节的间接照明系统，减少多达 60% 的室内眩光以及能源消耗；

• 特大双层低辐射玻璃幕墙确保自然光穿透，同时减少能源消耗，并有助于隔热以及隔音；

• 采用先进的光学催化技术 / 纳米光子净化系统，以改善室内空气质量；

• 中水回收处理系统可将处理后的中水用作厕所用水；

• 二氧化碳感应器可调控室内鲜风供应，确保空气清新怡人，同时避免能源过耗；

这个景观公园也作为一个绿化屋顶，隔开了楼下的商务区。这是一个符合公共公园需要的可持续的景观设计方案。该平台公园在景观设计中有以下几个用途：雨水存留，楼体保温，降低能耗，提高空气质量，培养原生态生活习惯，降低城市空气温度，减少热岛效应。

这些体贴细腻的设计策略反过来提高了该开发项目的能量绩效。还有多种方法齐头并进，如采用当地树木和植被、使用循环材料、建设废品管理站等。这些都对良好的可持续景观设计实践具有促进作用。设计理念体现了在建筑和城市的范畴内、不同景观间的互动。

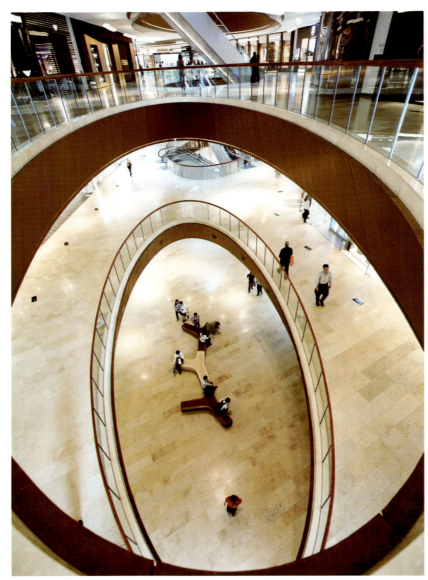

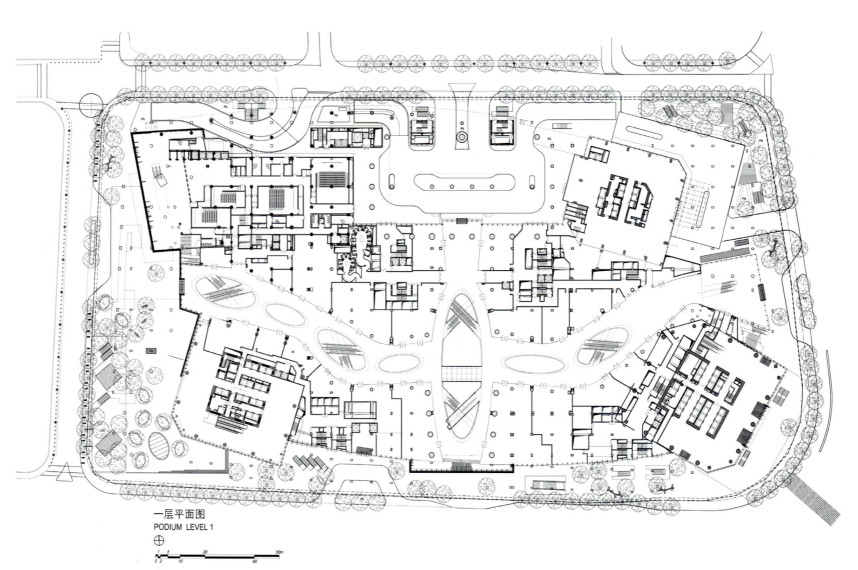

一层平面图
PODIUM LEVEL 1

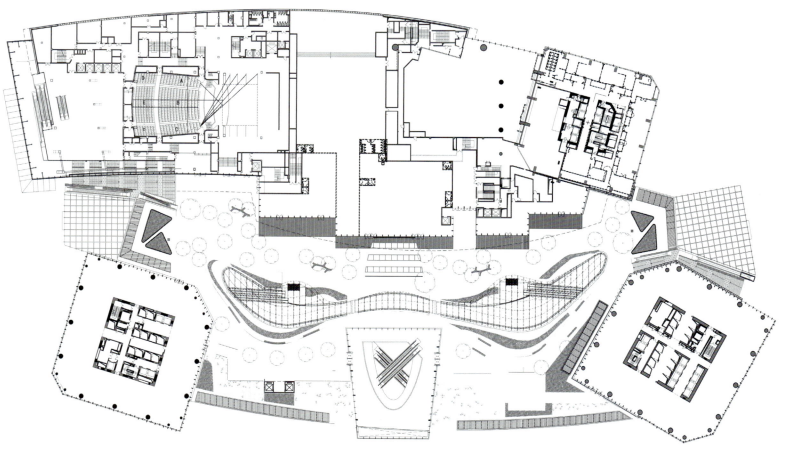

三层平面图
PODIUM LEVEL 3

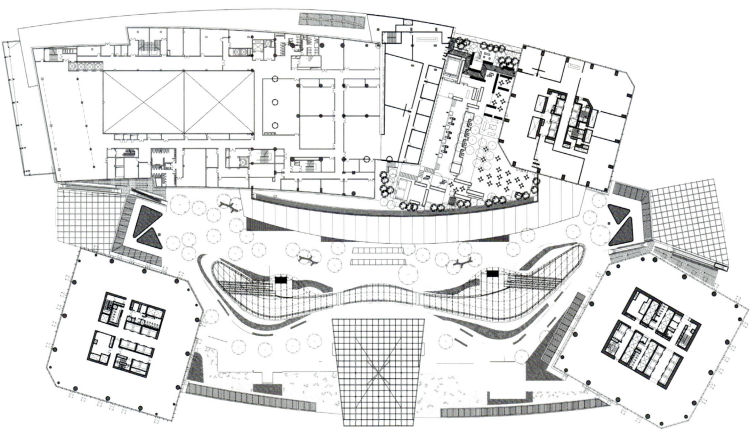

四层平面图

PODIUM LEVEL 4

9 SQUARE SHOPPING CENTER 九方购物中心

ARCHITECT
RTKL Associates Inc.

INTERIORS
Kevin Horn, Yuwen Peng, Xi Ren, Eric Yeh, Benny Chou

LOCATION
Chengdu, Sichuan Province, China

EXTERIORS
Norm Garden, David Schmitz, Kaz Miwa, Giselle Leung, Liang Zhong, Joanna Wong, Loni Valle

CLIENT
Shenzhen CATIC Real Estate Development Co., Ltd.

AREA
78,000 m²

PHOTOGRAPHER
RTKL/David Whitcomb

SUSTAINABLE & GREEN FEATURES 绿色特征

• *Use of hollow Low-E glass curtain wall* ／ 使用中空 Low-E 玻璃幕墙

• *Household metering central air-conditioning system* ／ 分户计量的中央空调系统

• *High isolation and noise insulation* ／ 高度隔热、隔音

Taking advantage of its location within the civic district and adjacent to a city park, CATIC Plaza in Chengdu presents two different façades – a bolder, streamlined elevation facing the primary roadway and a layered, varied elevation opening onto outdoor plazas connecting to the park. While the design of the building is contemporary and efficient, the team also wanted to reflect the city's rich heritage through this 78,000 m^2 retail, entertainment and office project. Special exterior perforated metal panels were created to celebrate Chengdu's famous woven brocades. The panels incorporate patterns found in these traditional weavings and create a subtle texture on the façade during the day; at night, they are back-lit with changing color to evoke a dramatic, sophisticated atmosphere.

This reference to the Chengdu brocade is continued and explored throughout the interiors as well. The layout is comprised of an elliptical center court with two legs on either side, while the ceiling of the curved side of the mall is highlighted by a sweeping band of patterned punched and back-lit metal, recalling similar details to the building's façade. Countering the glass handrail, the straight side of the mall is laminated with colored bands of translucent silk fabric that flow up through the space, connecting the various floors and highlighting the form and materials of the brocade. In the center court, crisscrossing lines of light and fiber optics run vertically along the bulkhead and create a loose web of woven threads to speak to the structure and the craft of making the brocades. Additional detail of pattern and form from the underside of the escalators to the vanities in the restroom speak to the flowing form and texture of brocade fabric.

东立面图 EAST ELEVATION

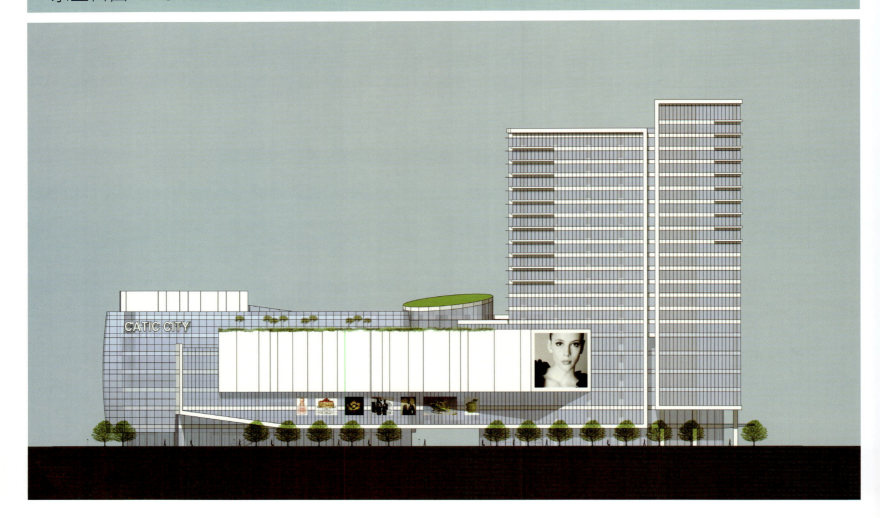

成都中航城市广场坐落在行政区内，毗连城市公园，地理位置优越。该广场呈现出两种不同的外观，面朝主干道的一面轮廓鲜明、简洁朴实，而通往露天广场、连接城市公园的一面则具有多层次、多样态的特点。项目面积为78000平方米，集零售、娱乐和办公于一身，既现代又高效。尽管如此，设计团队依旧想透过它展现其所在城市的丰富的文化遗产。成都蜀锦举世闻名，设计团队在建筑外观采用多孔金属板来弘扬成都的丝织品文化。这种金属板采用了传统丝织品所使用的图案，使建筑外观在白天呈现出淡雅之感；到了晚上，这些金属板则充当幕布，上面变换的色彩营造出一种引人注目、令人迷幻的氛围。

该项目在设计外观上对成都蜀锦的借鉴也沿用到了其内部设计中。内部由一个椭圆形中心广场和其两边的两个分支共同组成，购物中心弧形边的天花板在大幅度弯曲的穿孔背光金属边的映衬下尤为突出，令人回想起建筑外观上相似的设计手法。与玻璃扶手相对，购物中心的直边一面采用贯穿整个空间的半透明丝绸彩色金属边进行分层，连接多个楼层，使蜀锦的形态和材料更加突出。在中央广场上，灯光和光纤交织在一起，沿着防水壁垂直而下，形成一张松散、由编织线构成的网，表明了其结构及蜀锦的制作工艺。除此之外，从自动扶梯下方到洗手间内洗面台的图案和形式的设计也体现了丝织品流动的形态与质地。

南立面图 SOUTH ELEVATION

北立面图 NORTH ELEVATION

西立面图 WEST ELEVATION

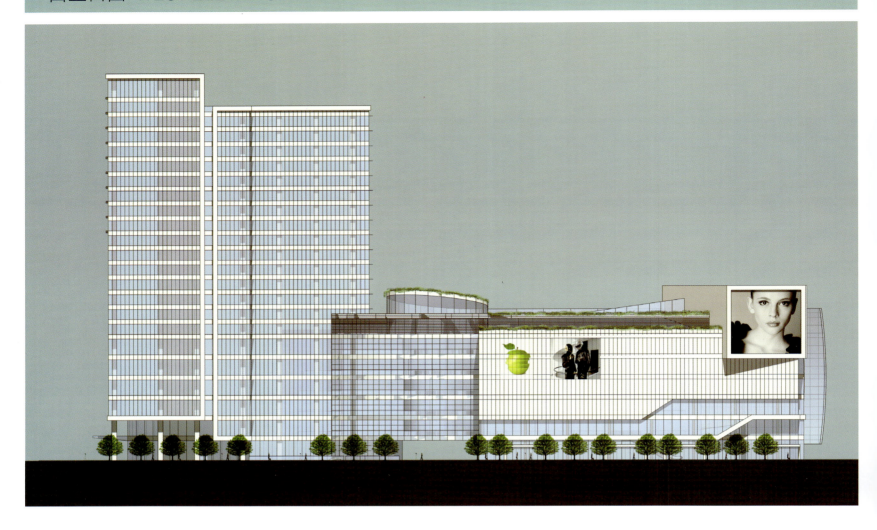

剖面图1 SECTION 1

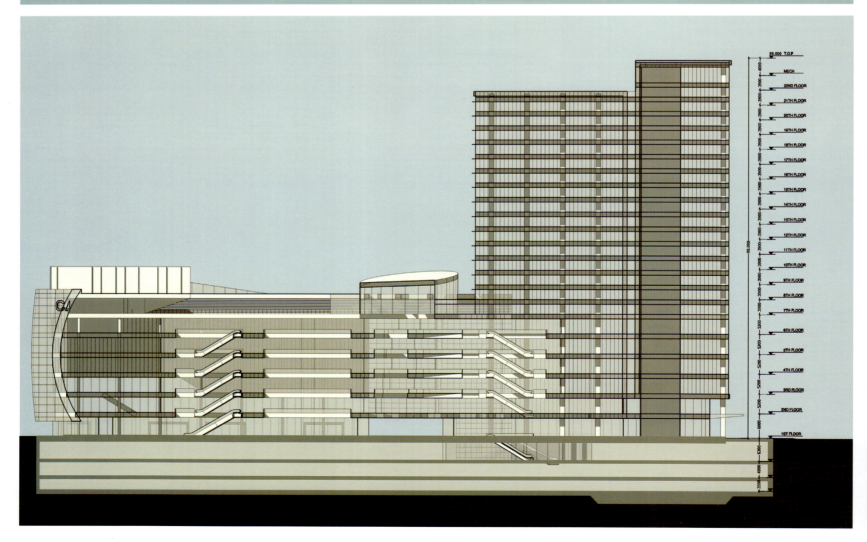

剖面图 2 SECTION 2

总平面图 / 屋顶平面图 SITE/ROOF PLAN

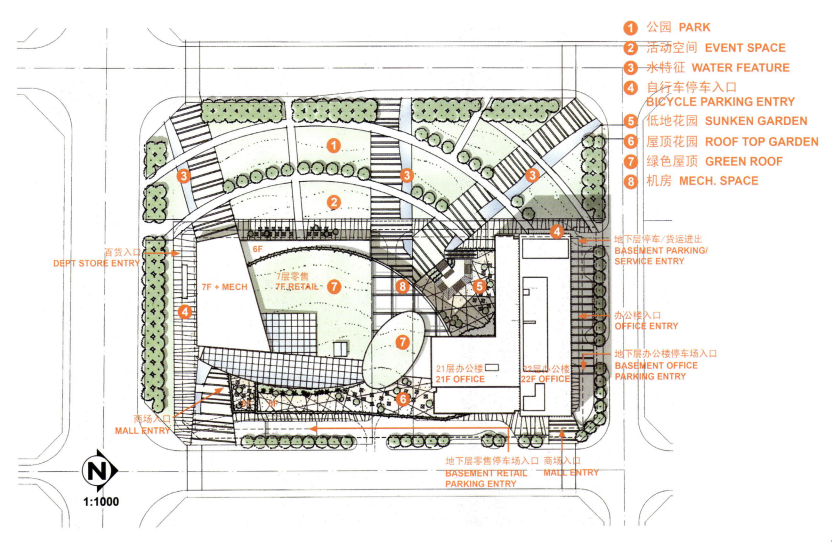

1 公园 PARK
2 活动空间 EVENT SPACE
3 水特征 WATER FEATURE
4 自行车停车入口 BICYCLE PARKING ENTRY
5 低地花园 SUNKEN GARDEN
6 屋顶花园 ROOF TOP GARDEN
7 绿色屋顶 GREEN ROOF
8 机房 MECH. SPACE

百货入口 DEPT STORE ENTRY

商场入口 MALL ENTRY

地下层停车/货运进出 BASEMENT PARKING/SERVICE ENTRY

办公楼入口 OFFICE ENTRY

地下层办公楼停车场入口 BASEMENT OFFICE PARKING ENTRY

地下层零售停车场入口 BASEMENT RETAIL PARKING ENTRY

商场入口 MALL ENTRY

6F

7F + MECH

7层零售 7F RETAIL

21层办公楼 21F OFFICE

22层办公楼 22F OFFICE

1:1000

B3 PLAN

商业 RETAIL	百货 DEPT. STORE	运动/娱乐 SPORT/ENTERTAINMENT
服务后勤 BACK OF HOUSE	餐饮娱乐 F&B	户外空间 OUTDOOR SPACE
走廊 COMMON AREA	办公室 OFFICE	
次要商业 JR. ANCHOR	核心筒,交通 VERTICAL CIRCULATION	

B2 PLAN

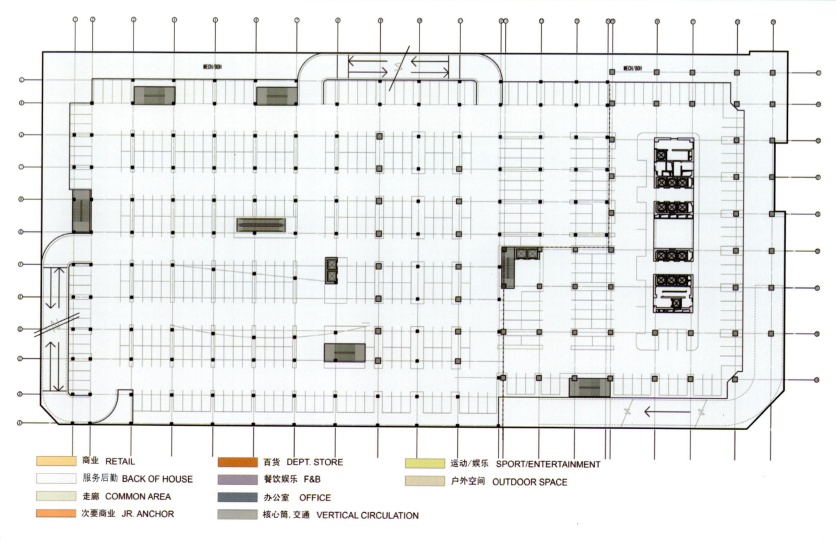

商业 RETAIL	百货 DEPT. STORE	运动/娱乐 SPORT/ENTERTAINMENT
服务后勤 BACK OF HOUSE	餐饮娱乐 F&B	户外空间 OUTDOOR SPACE
走廊 COMMON AREA	办公室 OFFICE	
次要商业 JR. ANCHOR	核心筒,交通 VERTICAL CIRCULATION	

B1 PLAN

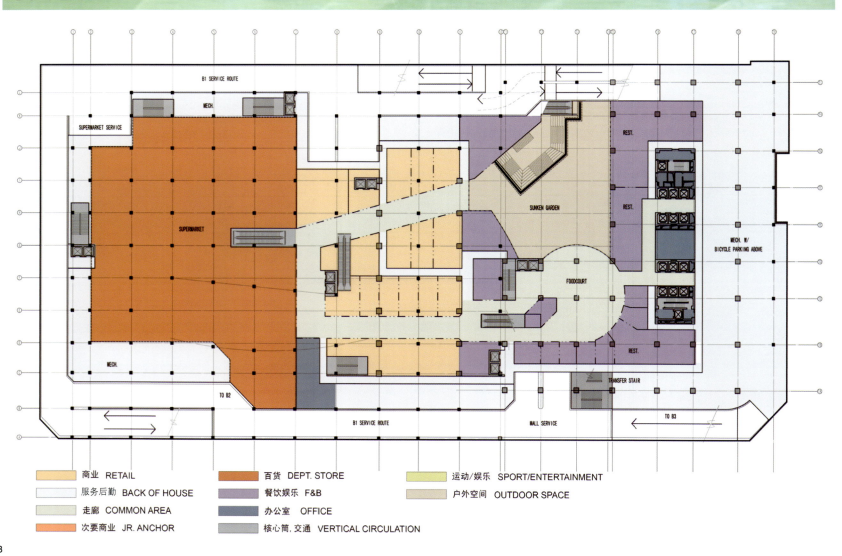

商业 RETAIL　　　　　　　　百货 DEPT. STORE　　　　　　　运动/娱乐 SPORT/ENTERTAINMENT

服务后勤 BACK OF HOUSE　　　餐饮娱乐 F&B　　　　　　　　户外空间 OUTDOOR SPACE

走廊 COMMON AREA　　　　　　办公室 OFFICE

次要商业 JR. ANCHOR　　　　　核心筒, 交通 VERTICAL CIRCULATION

L1 PLAN

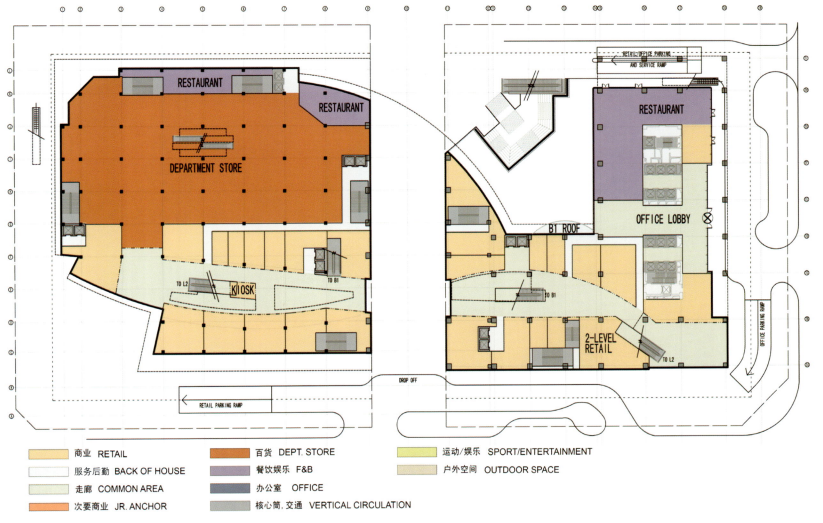

	商业 RETAIL		百货 DEPT. STORE		运动/娱乐 SPORT/ENTERTAINMENT
	服务后勤 BACK OF HOUSE		餐饮娱乐 F&B		户外空间 OUTDOOR SPACE
	走廊 COMMON AREA		办公室 OFFICE		
	次要商业 JR. ANCHOR		核心筒, 交通 VERTICAL CIRCULATION		

L2 PLAN

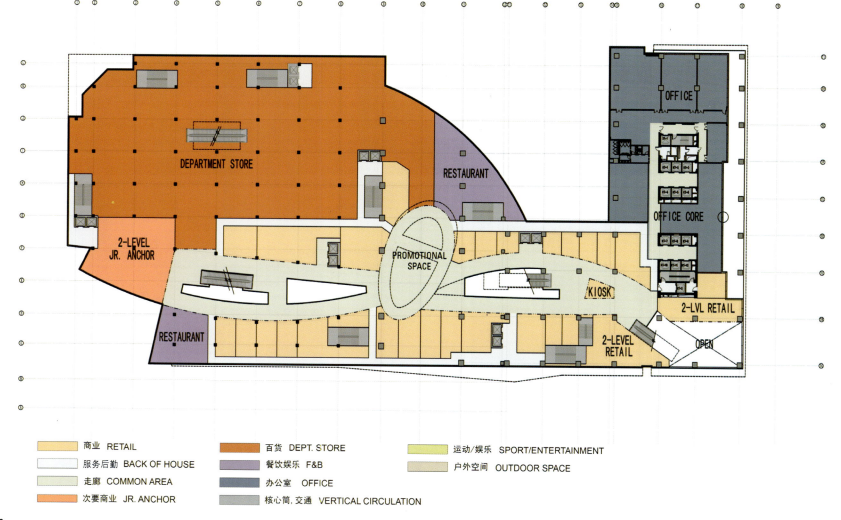

OFFICE

OFFICE CORE

DEPARTMENT STORE

RESTAURANT

2-LEVEL JR. ANCHOR

PROMOTIONAL SPACE

KIOSK

2-LVL RETAIL

RESTAURANT

2-LEVEL RETAIL

OPEN

	商业 RETAIL		百货 DEPT. STORE		运动/娱乐 SPORT/ENTERTAINMENT
	服务后勤 BACK OF HOUSE		餐饮娱乐 F&B		户外空间 OUTDOOR SPACE
	走廊 COMMON AREA		办公室 OFFICE		
	次要商业 JR. ANCHOR		核心筒,交通 VERTICAL CIRCULATION		

L3 PLAN

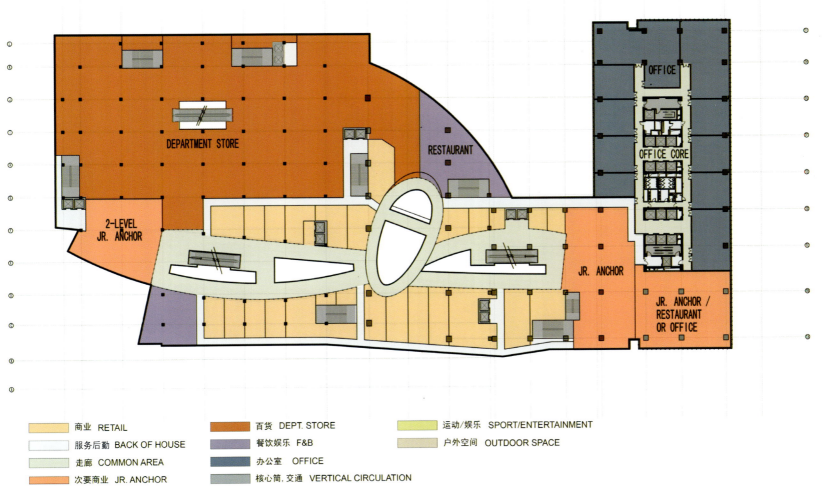

DEPARTMENT STORE

RESTAURANT

OFFICE

OFFICE CORE

2-LEVEL JR. ANCHOR

JR. ANCHOR

JR. ANCHOR / RESTAURANT OR OFFICE

	商业 RETAIL		百货 DEPT. STORE		运动/娱乐 SPORT/ENTERTAINMENT
	服务后勤 BACK OF HOUSE		餐饮娱乐 F&B		户外空间 OUTDOOR SPACE
	走廊 COMMON AREA		办公室 OFFICE		
	次要商业 JR. ANCHOR		核心筒,交通 VERTICAL CIRCULATION		

L4 PLAN

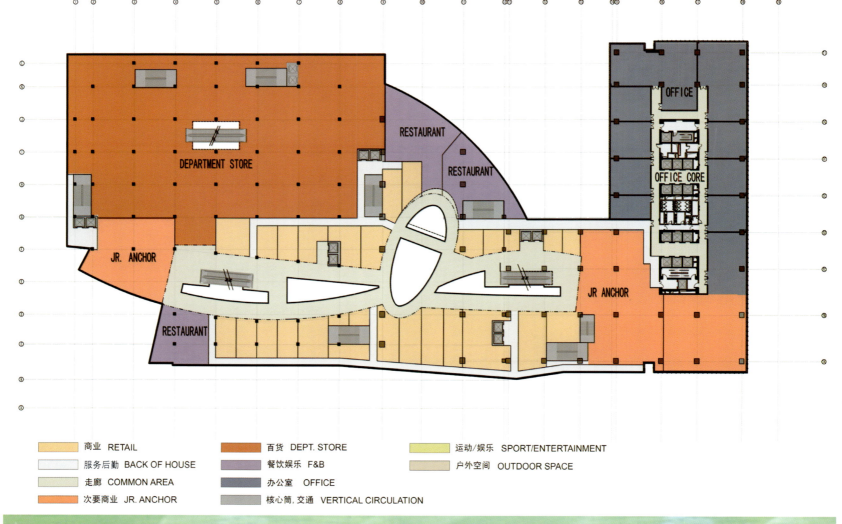

商业 RETAIL	百货 DEPT. STORE	运动/娱乐 SPORT/ENTERTAINMENT
服务后勤 BACK OF HOUSE	餐饮娱乐 F&B	户外空间 OUTDOOR SPACE
走廊 COMMON AREA	办公室 OFFICE	
次要商业 JR. ANCHOR	核心筒,交通 VERTICAL CIRCULATION	

L5 PLAN

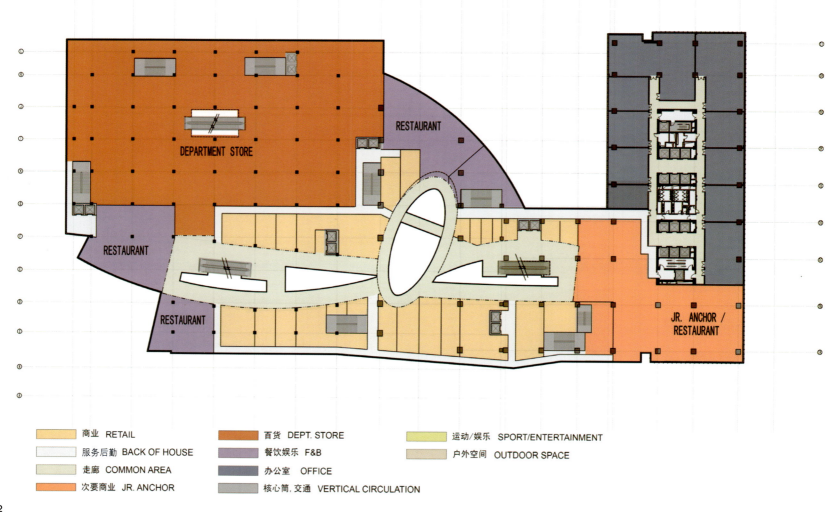

商业 RETAIL	百货 DEPT. STORE	运动/娱乐 SPORT/ENTERTAINMENT
服务后勤 BACK OF HOUSE	餐饮娱乐 F&B	户外空间 OUTDOOR SPACE
走廊 COMMON AREA	办公室 OFFICE	
次要商业 JR. ANCHOR	核心筒,交通 VERTICAL CIRCULATION	

CHINA RESOURCES BUILDING

华润大厦

2012 THE AMERICAN INSTITUTE OF HONG KONG ARCHITECTS HONG KONG CHAPTER – SUSTAINABLE DESIGN FOR ARCHITECTURE　2012 年度美国建筑师学会香港分会建筑可持续设计奖

ARCHITECT
Ronald Lu & Partners

LOCATION
Hong Kong, China

AREA
6,600 m² (Site Area), 99,000 m² (Built Area)

SUSTAINABLE & GREEN FEATURES　绿色特征

- Highly efficient glass curtain wall to reduce air-conditioning load as well as energy consumption ╱ 高效玻璃幕墙减低空调负荷，节省能源
- Innovative and highly efficient lighting system and CO_2 requirement controlled ventilation system ╱ 创新的高效照明系统和二氧化碳需求控制通风系统

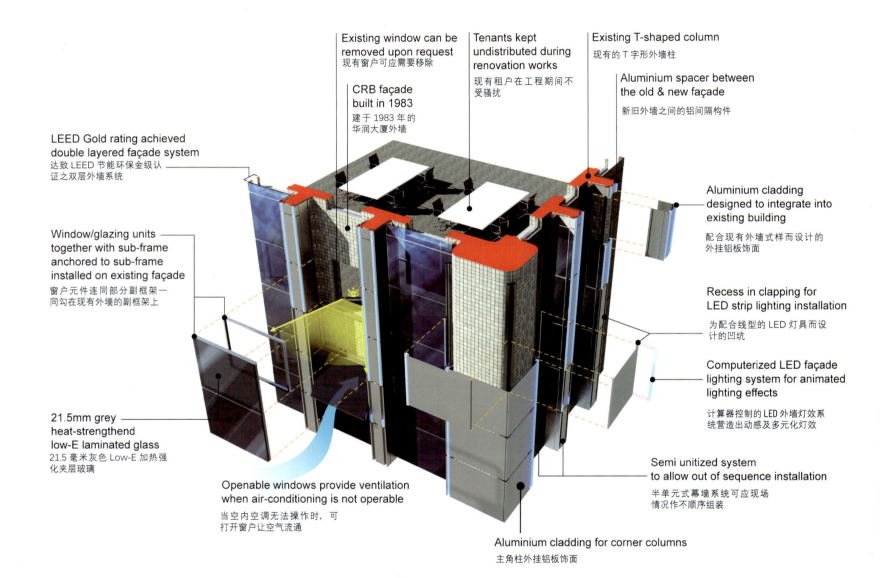

LEED Gold rating achieved double layered façade system
达致 LEED 节能环保金级认证之双层外墙系统

Window/glazing units together with sub-frame anchored to sub-frame installed on existing façade
窗户元件连同部分副框架一同勾在现有外墙的副框架上

21.5mm grey heat-strengthend low-E laminated glass
21.5 毫米灰色 Low-E 加热强化夹层玻璃

Openable windows provide ventilation when air-conditioning is not operable
当空内空调无法操作时，可打开窗户让空气流通

Existing window can be removed upon request
现有窗户可应需要移除

CRB façade built in 1983
建于 1983 年的华润大厦外墙

Tenants kept undistributed during renovation works
现有租户在工程期间不受骚扰

Existing T-shaped column
现有的 T 字形外墙柱

Aluminium spacer between the old & new façade
新旧外墙之间的铝间构件

Aluminium cladding designed to integrate into existing building
配合现有外墙式样而设计的外挂铝板饰面

Recess in clapping for LED strip lighting installation
为配合线型的 LED 灯具而设计的凹坑

Computerized LED façade lighting system for animated lighting effects
计算器控制的 LED 外墙灯效系统营造出动感及多元化灯效

Semi unitized system to allow out of sequence installation
半单元式幕墙系统可应现场情况作不顺序组装

Aluminium cladding for corner columns
主角柱外挂铝板饰面

Constructed in the 1980s, China Resources Building, was a landmark in Wan Chai District. Its owner decided to launch the renovation project of the headquarters so as to coordinate with the commercial development activities of the northern part of Wan Chai District and promote corporate images.

China Resources Building is the first renovation project that gets in advance the Leadership in Energy and Environmental Design (LEED) gold certification of the United States Green Building Council. After landscaping, operating expenses and energy can be reduced, and the quality of the air can be improved and tenents will be provided with better indoor environment. In the whole renovation process, design and construction play an important role.

Methods of Construction

Up to 95% of the existing walls, floors and roofs are kept to prolong the life cycle of the building. Construction waste management plans to recycle at least 50% safe construction and demolition waste. The material of the new exterior wall is produced in areas nearby, which can decrease the emission of carbon during the process of transportation.

High Performance Exterior Wall

The renovation project transforms the curtain wall of the building into glass curtain wall that can reduce the absorption of solar energy. Only

5% solar energy can penetrate into the indoor space, which can cut down air conditioning. In order to achieve the long-term goal of energy conservation, the exterior wall introduces vertical LED lighting system as well as the program control, so that the façade lighting can change flexibly according to varied occasions.

Energy Conservation and Reduction of Greenhouse Gas Emission

The interior of the building is also equipped with an innovative lighting system. Integrated with the technology of sun sensor and flow sensor, the lighting will automatically go out when exceeding the preprogrammed time to cut down the cost. A series of measures and various relevant equipments are seen so as to encourage users to recycle and reuse the waste.

Improvement of Indoor Air Quality

Ventilating system is installed to keep the indoor carbon dioxide standard and change the volume of indoor air accordingly. When moderate natural wind is brought into the building, the air quality will be improved and thus energy saved. The interior decoration materials, including oil paint, carpet, tackifier and aquaseal are eco-friendly, and the organic compound materials are of low volatility. All these guarantee a reduced indoor air pollution, as well as a more healthy and comfortable environment for the tenants.

华润大厦建于 20 世纪 80 年代,是当时湾仔区的地标性大厦。为配合湾仔北区的商业发展活动,以及提高企业形象,业主决定开展总部翻新工程。

华润大厦是首个预先得到美国绿色建筑协会的能源与环境设计先锋(LEED)金奖认证的翻新工程。绿化后不但减少了营运成本,更加节能,亦改善了空气质量,同时也为租户提供了更好的室内环境。在整个翻新过程中,设计与施工同样重要。

施工方法

项目保留原有的墙、地板和屋顶,比率高达 95%,延长了建筑的生命周期。建筑废物管理计划回收至少 50% 的非危险建筑和拆建废料。新外墙的材料由就近地区生产,可减少在运输过程中产生的碳排放。

高效能外墙

翻新工程将大厦幕墙改为可减少吸收太阳能的玻璃幕墙。只让 5% 的太阳能透入室内,进而达到减低空调负荷的目的。为达到长远节能的目标,设计于外墙引入垂直的 LED 灯光系统,配以程控,使外墙灯光效果可应不同场合而灵活变化。

节能及减低温室气体排放

大厦内亦设有一个创新的照明系统,结合了日光传感器和人流传感器的技术,灯光会在超过预设时间的空置时间内自动熄灭,以减低照明成本。大厦还推行了一系列措施并安装了各项有关设备,以鼓励用户将废物回收及循环再用。

改善室内空气质量

室内安装了通风系统,以保持二氧化碳在室内的水平,进而改变室内空气总量。通过引入适量的自然风,达到改善空气质量和节能的效果。在选择室内装修材料时,包括油漆、地毯、胶黏剂和密封剂等,都选用了低挥发性的有机化合物材料,这不但可减少室内的空气污染,还能为各用户提供一个更健康及舒适的环境。

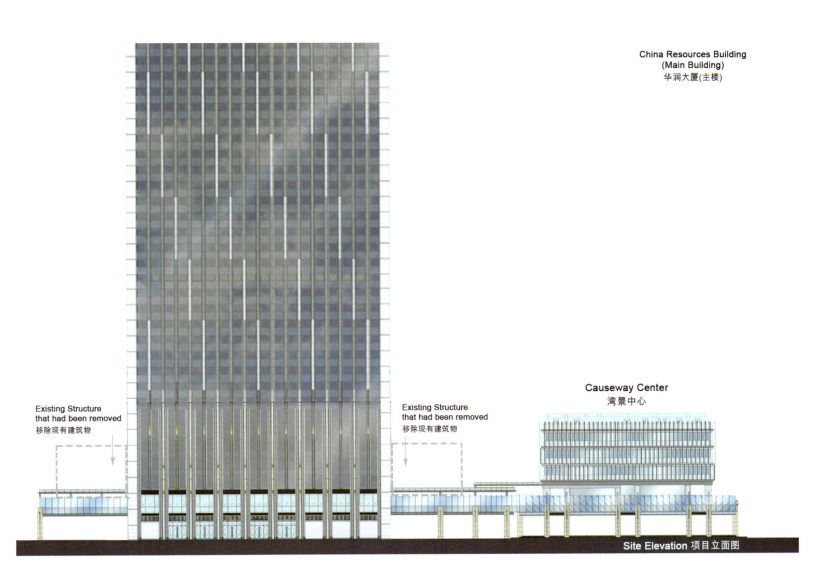

China Resources Building
(Main Building)
华润大厦(主楼)

Existing Structure
that had been removed
移除现有建筑物

Existing Structure
that had been removed
移除现有建筑物

Causeway Center
湾景中心

Site Elevation 项目立面图

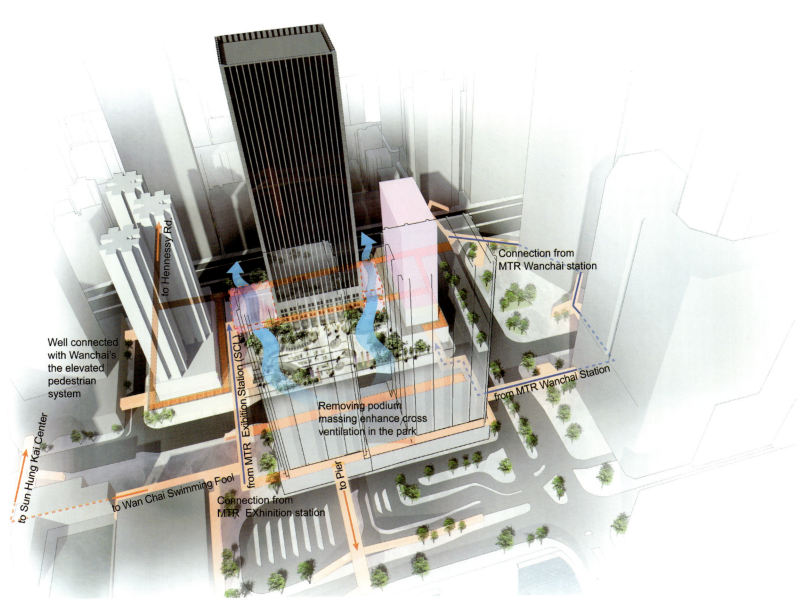

to Hennessy Rd.

Connection from
MTR Wanchai station

Well connected
with Wanchai's
the elevated
pedestrian
system

from MTR Wanchai Station

from MTR Exhibition Station (SCL)

to Sun Hung Kai Center

to Wan Chai Swimming Fool

to Pier

Removing podium
massing enhance cross
ventilation in the park

Connection from
MTR EXhinition station

Future Hotel Development
未来酒店项目

Harbour Road Garden
港湾道公园

Causeway Centre
湾景中心

China Resources Building
(Main Building)
华润大厦(主楼)

Site Plan

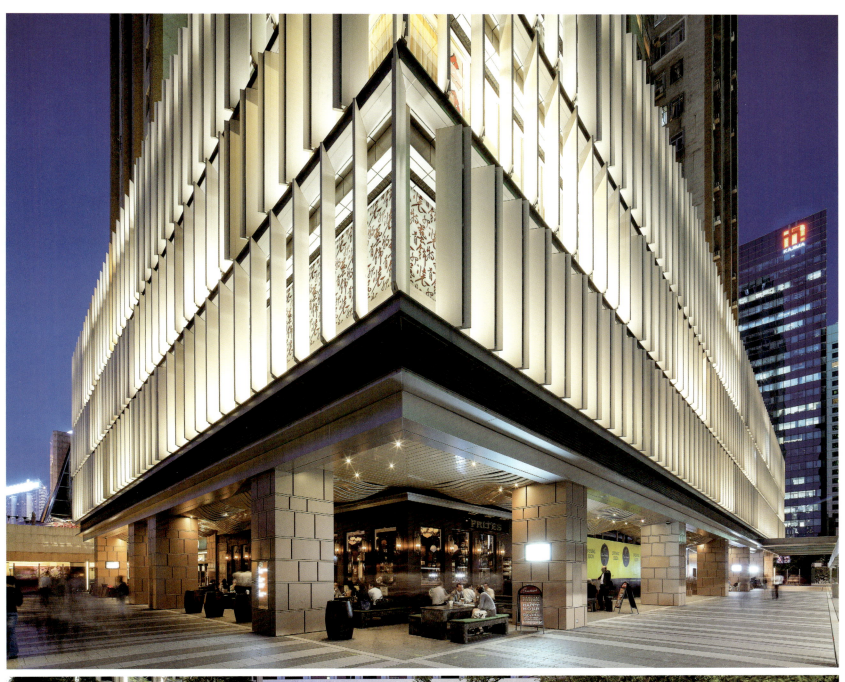

SIU SAI WAN COMPLEX

小西湾综合大楼

2012 GREEN BUILDING AWARD GRAND AWARD　2012 年度绿色建筑优秀奖

2011 THE AMERICAN INSTITUTE OF ARCHITECTS HK, SUSTAINABLE DESIGN AWARD
2011 年度美国建筑师学会香港分会建筑可持续设计奖

ARCHITECT
Ronald Lu & Partners

LOCATION
Hong Kong, China

AREA
4,400 m² (Site Area), 25,923 m² (Built Area)

SUSTAINABLE & GREEN FEATURES　绿色特征

- *Green roof irrigated with grey water/rainwater recycling system* ／ 绿色屋顶由处理过的雨水、污水灌溉
- *Use of vertical sun-shading fins, operable windows, insulated low-E glass and solar vacuum tubes* ／ 使用垂直遮阳板、可开闭窗户、低辐射玻璃和太阳能真空管
- *Daylight-lit and natural ventilated central atrium* ／ 中庭采用自然光及自然通风
- *Pre-cool/pre-heat fresh air intake for arena* ／ 体育馆使用预先冷却和加热的新鲜空气
- *Horizontal slats for sun-shading and rain-screening* ／ 水平式百叶板用作遮阳及挡雨

Siu Sai Wan complex building features a 1000-seat multi-role indoor gym, two indoor pools, a library, a table tennis room, a playroom, two interlinked activity rooms and a community hall which can accommodate 450 people. As a green building, it provides leisure sports facilities for residents at Island East.

Open and diverse architecture designs promote interaction with communities, reduce energy consumption and lead residents inside of the building for activities. The sunshade curtain wall at the main entrance looks like trees, extending outward to the courtyard. Residents can enter the building through the big stairs at street entrance or the pedestrian overpass which connects the mall across the street and the bus terminus.

Courtyard designed as "Vertical Street" locates at the center of complex building with corridor, platform, bridge, stairs and elevator stretching up in a circle, leading to activity centers at each floor on both sides of the building. "Vertical Street" also offers the public a panoramic view of the surrounding. Different spaces in the courtyard may house all kinds of activities, encouraging interaction among the public and bringing vitality to the building.

The roof uses ground glass and horizontal louver boards on both open sides, allowing abundant sunshine and natural ventilation in the towering courtyard, thus reducing the energy consumption dramatically. The design also includes many green elements: automated windows that increase indoor natural ventilation, insulation energy-saving glass, external sunshade equipment, green roof, water-cooled clarification system, sewage circulation system, non-PVC carpet, water miscible oil paint, escalator with passenger sensor, ambient and light sensor, energy-saving blower and solar vacuum tubes that provide hot water.

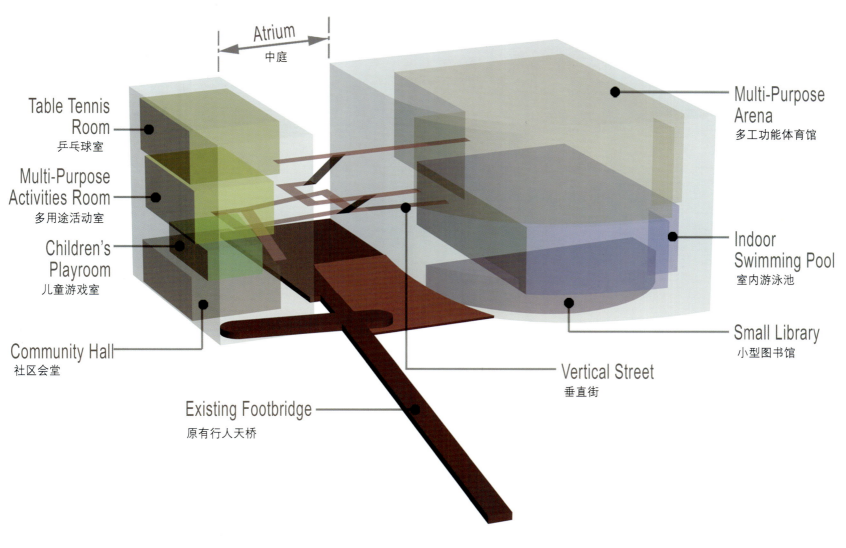

Table Tennis Room
乒乓球室

Multi-Purpose Activities Room
多用途活动室

Children's Playroom
儿童游戏室

Community Hall
社区会堂

Existing Footbridge
原有行人天桥

Atrium
中庭

Multi-Purpose Arena
多工能体育馆

Indoor Swimming Pool
室内游泳池

Small Library
小型图书馆

Vertical Street
垂直街

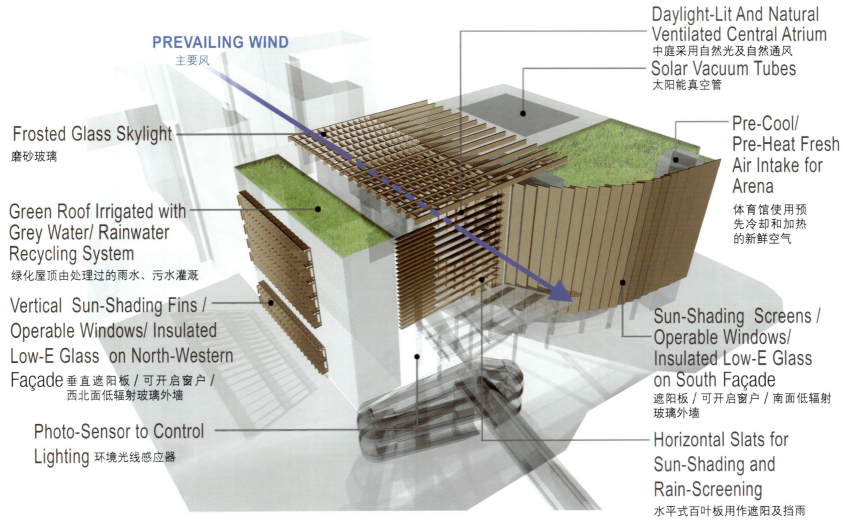

PREVAILING WIND
主要风

Frosted Glass Skylight
磨砂玻璃

Green Roof Irrigated with Grey Water/ Rainwater Recycling System
绿化屋顶由处理过的雨水、污水灌溉

Vertical Sun-Shading Fins / Operable Windows/ Insulated Low-E Glass on North-Western Façade 垂直遮阳板 / 可开启窗户 / 西北面低辐射玻璃外墙

Photo-Sensor to Control Lighting 环境光线感应器

Daylight-Lit And Natural Ventilated Central Atrium
中庭采用自然光及自然通风
Solar Vacuum Tubes
太阳能真空管

Pre-Cool/ Pre-Heat Fresh Air Intake for Arena
体育馆使用预先冷却和加热的新鲜空气

Sun-Shading Screens / Operable Windows/ Insulated Low-E Glass on South Façade
遮阳板 / 可开启窗户 / 南面低辐射玻璃外墙

Horizontal Slats for Sun-Shading and Rain-Screening
水平式百叶板用作遮阳及挡雨

现有学校
EXISTING SCHOOL

现有学校
EXISTING SCHOOL

绿色屋顶
GREEN ROOF

披璃天窗
GLAZED SKYLIGHT

现有人行桥
EXISTING FOOTBRIDGE

玻璃顶棚
GLAZED CANOPY

金属屋顶
METAL ROOF

绿色屋顶
GREEN ROOF

N

　　小西湾综合大楼设有一个 1000 座的多用途室内体育馆、两个室内游泳池、图书馆、乒乓球室、儿童游戏室、两个可相通的活动室及一个可容纳 450 人的小区会堂，以绿色建筑为港岛东居民提供休闲及康乐体育设施。

　　开放而多形态的建筑设计，在促进小区互动交流的同时减低能源消耗，又将居民的活动引进大楼内。正门如树木般的挡阳幕墙，由户外延伸到中庭。居民可从街道入口的大梯阶或从连接对面商场及巴士总站的行人天桥进入大楼。

　　设计为"垂直街"的中庭位于综合大楼的中心，有走廊、平台、桥、楼梯和电梯巧妙地环绕向上伸延，通往大楼两边各层的活动中心。通过"垂直街"，公众可饱览大楼周边的环境。中庭内的多个不同特质的空间可灵活地容纳各种活动，鼓励公众交流，为大楼带来无限活力。

　　大楼顶部采用磨砂玻璃，两边开放位置设有水平式百叶板，令高耸的中庭得到充沛的日光及自然通风，大大减少了能源消耗。大楼的设计亦包含很多绿化元素：自动化的窗户增加室内自然通风、隔热节能玻璃、外置遮阳设备、绿化屋顶、水冷式净化系统、污水循环系统、不含聚氯乙烯的地毯和水融性油漆及涂料、人流感应扶手电梯、环境光线光传感器、节能送风机和提供热水的太阳能真空管。

K11 ART MALL

K11 购物艺术中心

LEED GOLD CERTIFIED 美国绿色建筑认证金级认证

ARCHITECT	**LOCATION**	**AREA**	**PHOTOGRAPHER**
Kokaistudios	Shanghai, China	35,500 m²	Charlie Xia and K11

SUSTAINABLE & GREEN FEATURES 绿色特征

- *Use of hollow Low-E glass and highly-efficient insulating layer* ／ 使用中空 Low-E 玻璃和高效隔热层

- *Roof garden and green wall to reduce heat island effect* ／ 屋顶花园和绿化墙用以减轻热岛效应

- *Irrigation system optimizes water consumption* ／ 灌溉系统降低耗水量

- *Use of recyclable materials and accessible public transport* ／ 使用可回收材料和无障碍公共交通

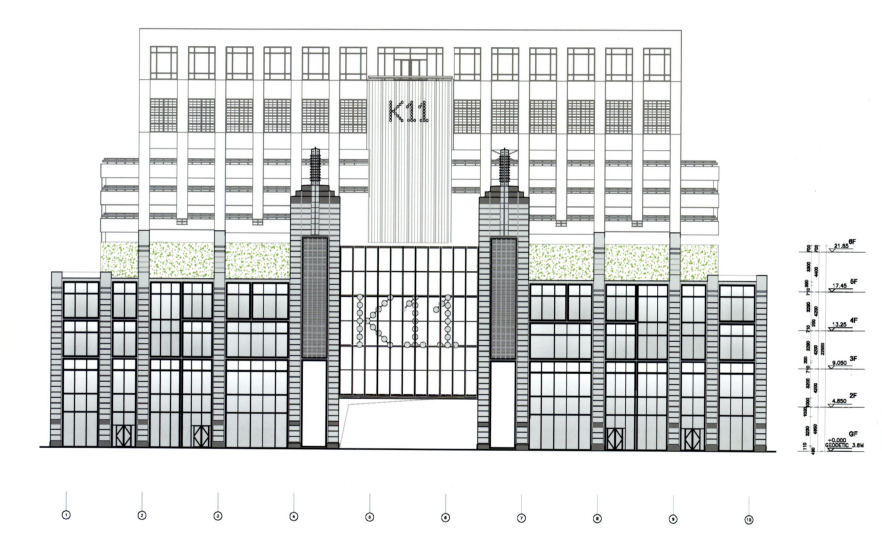

The transformation, encompassing architecture and interior design, of the 14,300 m² commercial podium of New World Tower in Huai Hai Road, into a "Lifestyle Center" strongly focused on "Art, People and Nature" opens up new scenarios for the commercial center of Shanghai. The redesign of the four main elevations of the building was conjugated in a way to balance opposite and sometime conflicting needs for conservation and innovation. The need to respect both the Huai Hai Road historical heritage and the New World Tower original design as dictated by the Shanghai municipal authorities needed to act in concert with the commercial needs of the developer to maximize visibility for the anchor tenants located at the lower two floors.

The "Core and Shell LEED Gold certification for Existing Buildings" has been applied for this project, and different strategies have driven our design in order to increase energy performance (Low-E Insulated Glass, High Performance Insulation), reduce the "heat island effect" (roof gardens and green walls), and reduce water use. In addition we have emphasized the use of sustainable materials throughout the project while also focusing on public transportation accessibility.

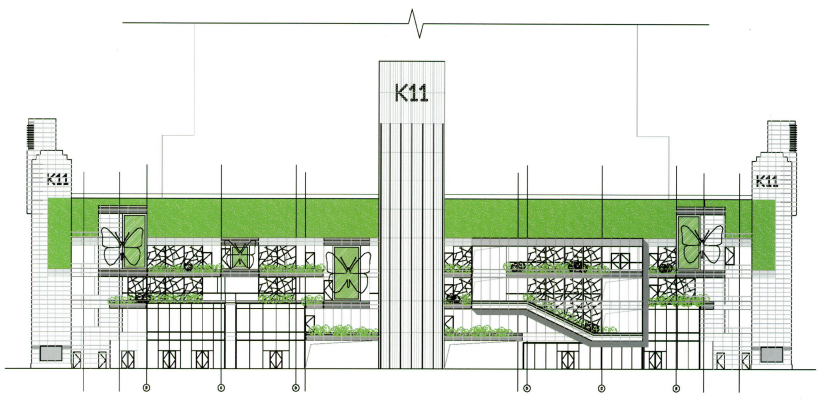

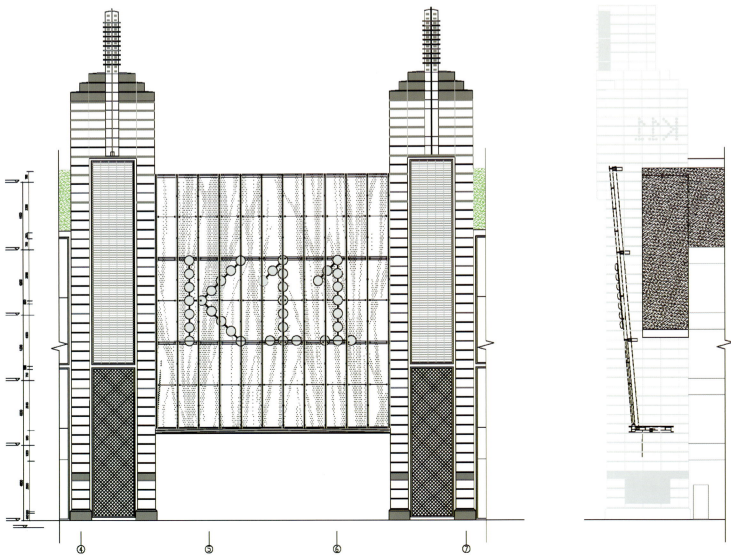

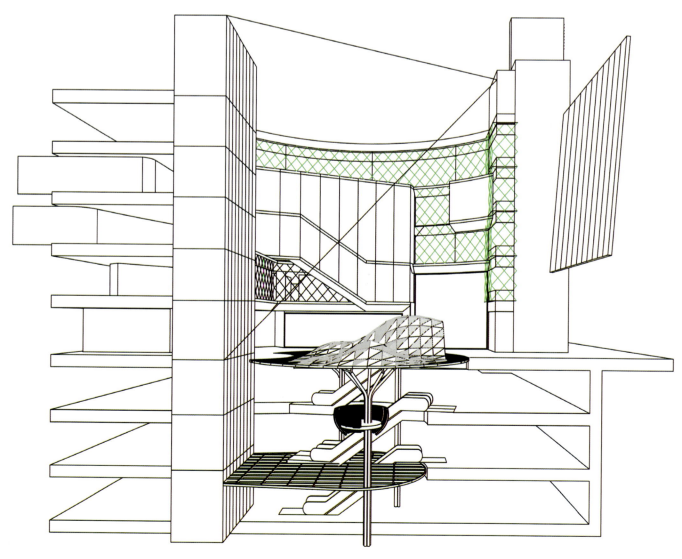

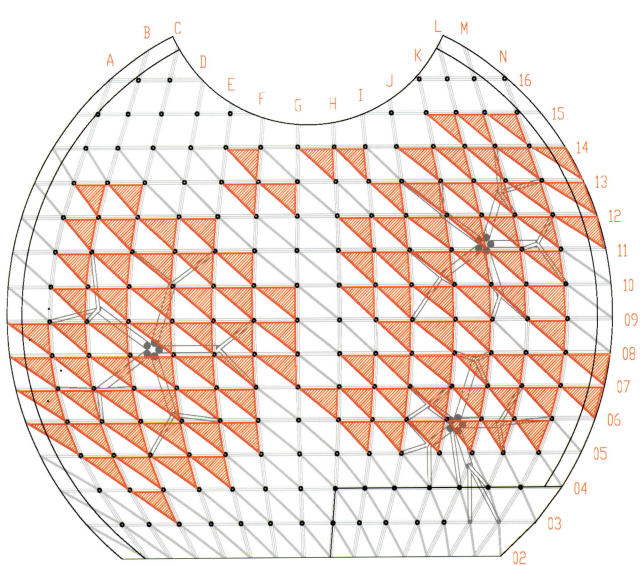

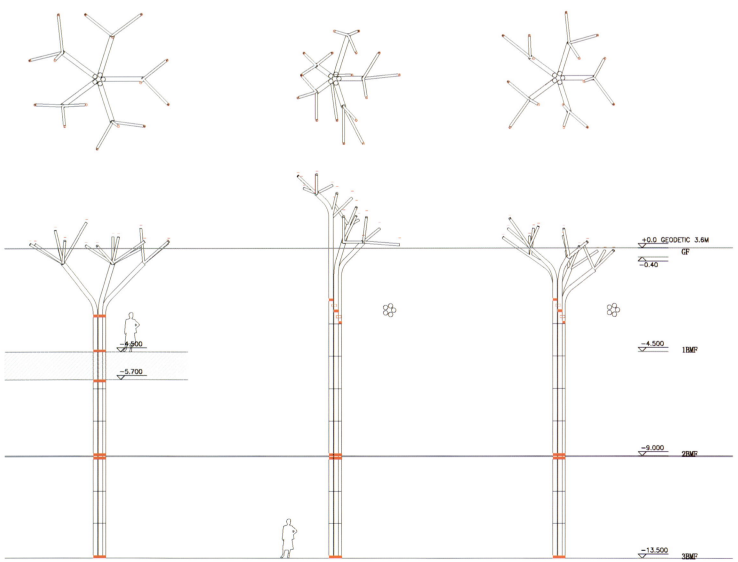

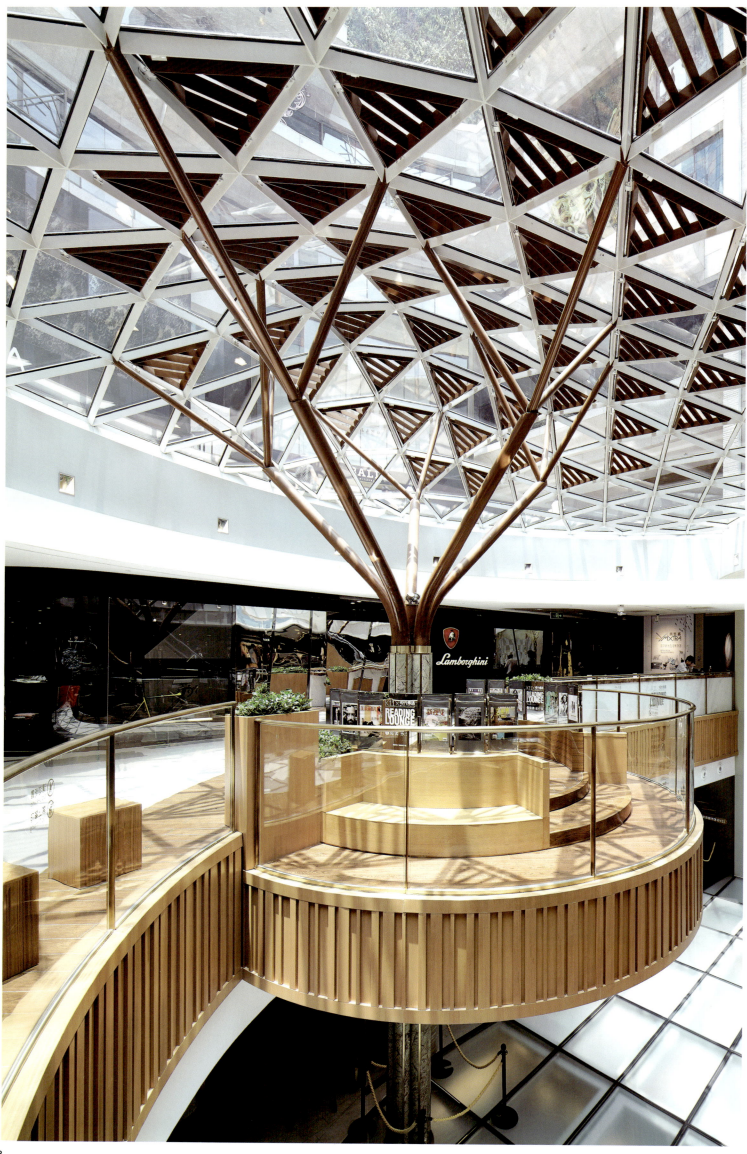

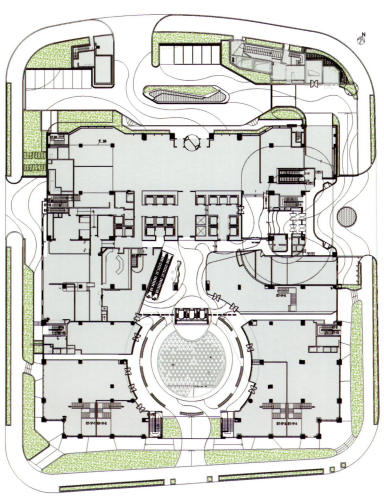

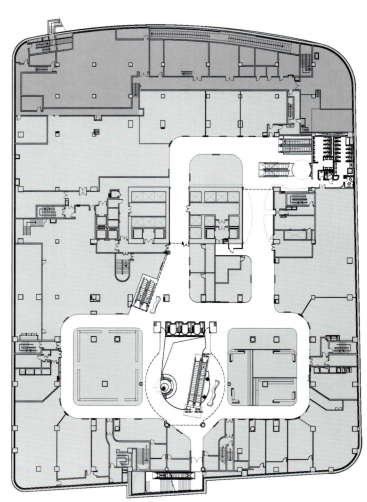

淮海路上的14300平方米的新
世纪塔楼裙房从建筑到室内转化成
全新的"生活中心"，秉持"艺术、
人文和自然"的信念为上海商业开
启新篇章。重新设计的四个主要立
面将守旧与创新这一对截然相反而
又时而冲突的元素平衡起来。一方
面是上海市政府对淮海路历史建筑
和新世界塔楼原样设计的重视。另
一方面，业主也希望为建筑底部一、
二层的商户营造出视觉的通透性。

此项目获得了"既有建筑环境
与设计先锋奖金项认证"，我们在
设计中还加入了不同的节能战略以
提高能源的使用效率（Low-E中空
玻璃、高性能的保温层），降低"热
岛效应"（屋顶花园和绿化墙），
减少用水量。同时，可持续性材料
的应用和无障碍公共交通的布局也
是此次设计的重点。

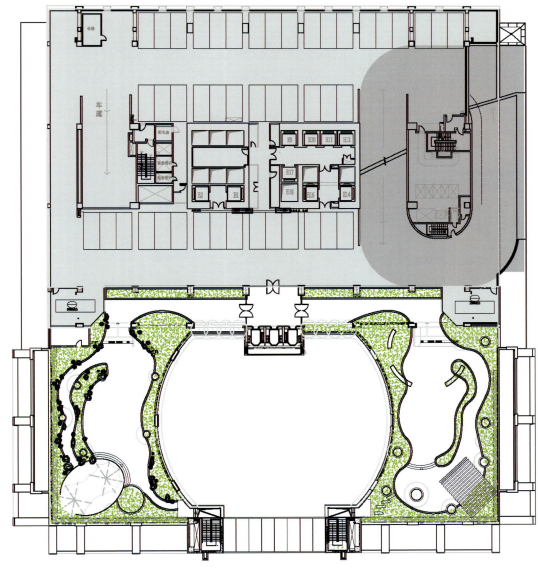

ZHANGJIANG LOTUS CREATIVE CENTER

上海张江集电港产业中心

ARCHITECT	LOCATION	AREA
JWDA	Shanghai, China	650,000 m²

SUSTAINABLE & GREEN FEATURES 绿色特征

- *Green roof to recycle rainwater and lighten heat island effect* ／ 绿色屋顶用于雨水回收，减轻热岛效应
- *Use of water-saving equipments* ／ 使用节水装置
- *Large use of daylighting and energy-saving lighting system and air conditioning system* ／ 大量使用自然采光、节能照明系统和空调系统
- *1+1 conditioning system to promote indoor climate and improve comfort* ／ 1+1 空调系统，改善室内气候，提高舒适性

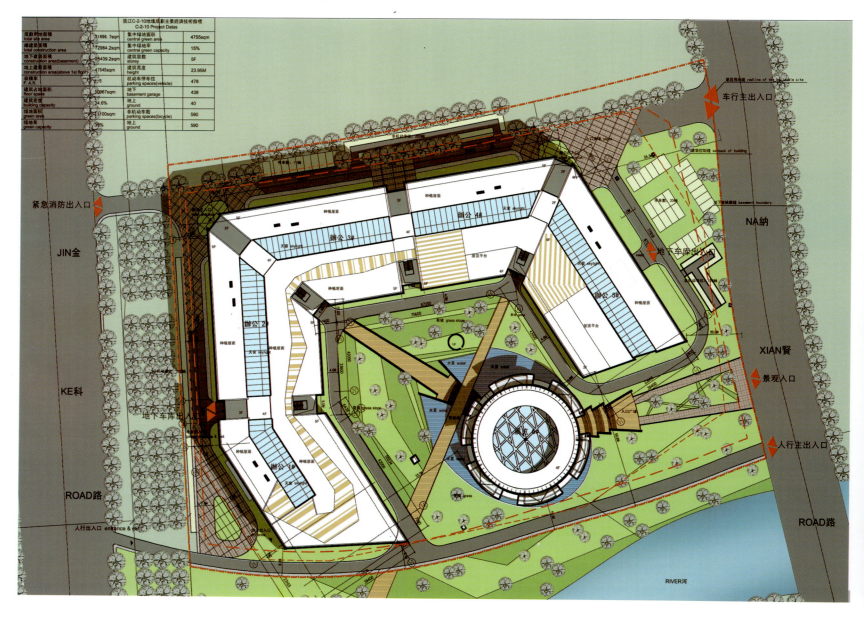

Four main design principles of this project are ecological harmony, efficiency and convenience, landscape rhythm, and openness and comfortableness. The design lays stress on the concept of ecology, communication and flexibility, making the creative center a colorful and friendly place. At the same time, it integrates convenient and practical supporting facilities and landscape activity space, striving to create harmony between users of the creative center and the public passing by and offer a work and life space which has an impact on and attraction for cultural and creative industries and the surrounding land of the creative center.

生态和谐、高效便捷、景观节奏和开放舒适是本项目的四个主要设计原则。在设计上突出了"生态性、交流性、灵活性"的理念，使创意园区呈现更加丰富和亲切的空间形态；同时，结合便捷实用的功能配套设施和景观活动场所，力求让园区使用者与过往人群和谐共生，为文化创意产业和周边地块创建具有影响力和吸引力的工作生活空间。

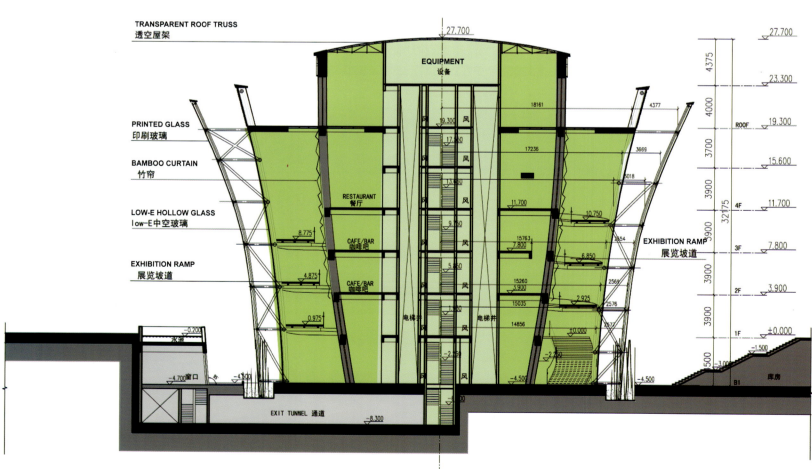

TRANSPARENT ROOF TRUSS
透空屋架

27.700

EQUIPMENT
设备

PRINTED GLASS
印刷玻璃

BAMBOO CURTAIN
竹帘

LOW-E HOLLOW GLASS
low-E中空玻璃

RESTAURANT
餐厅

EXHIBITION RAMP
展览坡道

CAFE/BAR
咖啡吧

CAFE/BAR
咖啡吧

电梯井 电梯井

风

水池

EXIT TUNNEL 通道

EXHIBITION RAMP
展览坡道

库房

地下层平面图 GROUND FLOOR PLAN

本层面积 ：25309m²

JIAYE INTERNATIONAL TOWN

嘉业国际城

ARCHITECT
9-town studio

LOCATION
Nanjing, Jiangsu Province, China

SUSTAINABLE & GREEN FEATURES 绿色特征

- *Special building structural design to maximize daylight and natural ventilation and avoid the direct sun* ／ 特殊建筑结构设计最大限度地引进自然光和自然通风，并避免日光直晒
- *Use of friendly and recyclable façade materials* ／ 表皮使用环保可回收材料

supported by the coexistence of pedestrians and vehicles.

The management of internal environment focuses on the quality of ecological environment. Yuantong Square is characterized by short depth work plane, which is convenient for direct ventilation and lighting. A study on the natural lighting is particularly made by designers. Hotel-office building adopts the corridor layout which is approximate symmetry. The north and south façades, integrated with the space for air conditioners, are designed with numerous horizontal components whose distribution modes are quite different because of the different needs of sunlight and skylight from the interior. In order to shelter from the strong sunlight in summer, especially the light rays at noon, the south façade sets the horizontal components of the air conditioning to the upper part of the indoor space. Layers of floors and other louvered horizontal components are also used as shelter from direct sunlight. The north façade, however, emphasizes the utilization of skylight. The horizontal components where air conditioners lie are in the lower part of the interior space, which forms a wide windowsill. They, together with other horizontal components, have the effect of refraction and diffusion of skylight. In this way, the interior space of the north gets abundant natural sunlight. In addition, the exterior façades are different in the north and the south: the south façade falls back within the horizontal components to reduce direct sunlight; the north façade protrudes, basically around the outside of horizontal components, in order to utilize more natural skylight.

The façade of the building comprises of green environmental friendly materials, such as red terracotta panel, bare concrete and glass curtain wall. Traditional plain terracotta panels shine with modern vibrant glass, like a dialogue between ancient and modern times, or that of nature and science. Such design brings cultural and ecological connotation to the entire central business district. Curtain wall made of terracotta panels uses terracotta panels of 210mm×1530mm and 305mm×315mm, as well as the correspondent terracotta shutter. The design combines big construction volume and fine texture decoration, and emphasizes on a more exquisite process design and work.

Sound functional layout and traffic management are top priorities in the design. The pedestrian space is emphasized and vehicle flow lines are set within the specified range. Motor vehicles use the entrance from the auxiliary street, and the traffic flow is moderated by the courtyard floor and the road type design. Thus the separation of vehicles from pedestrians guarantees the smooth traffic. The direct access of the distinguished guests to the entrance can avoid the traffic jam caused by vehicles. The environment is upgraded thanks to a efficient traffic organization with pedestrian traffic as the main

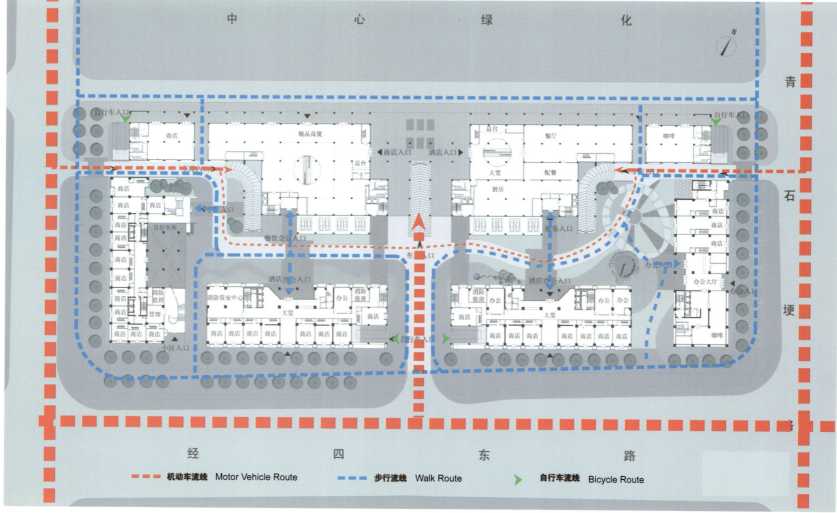

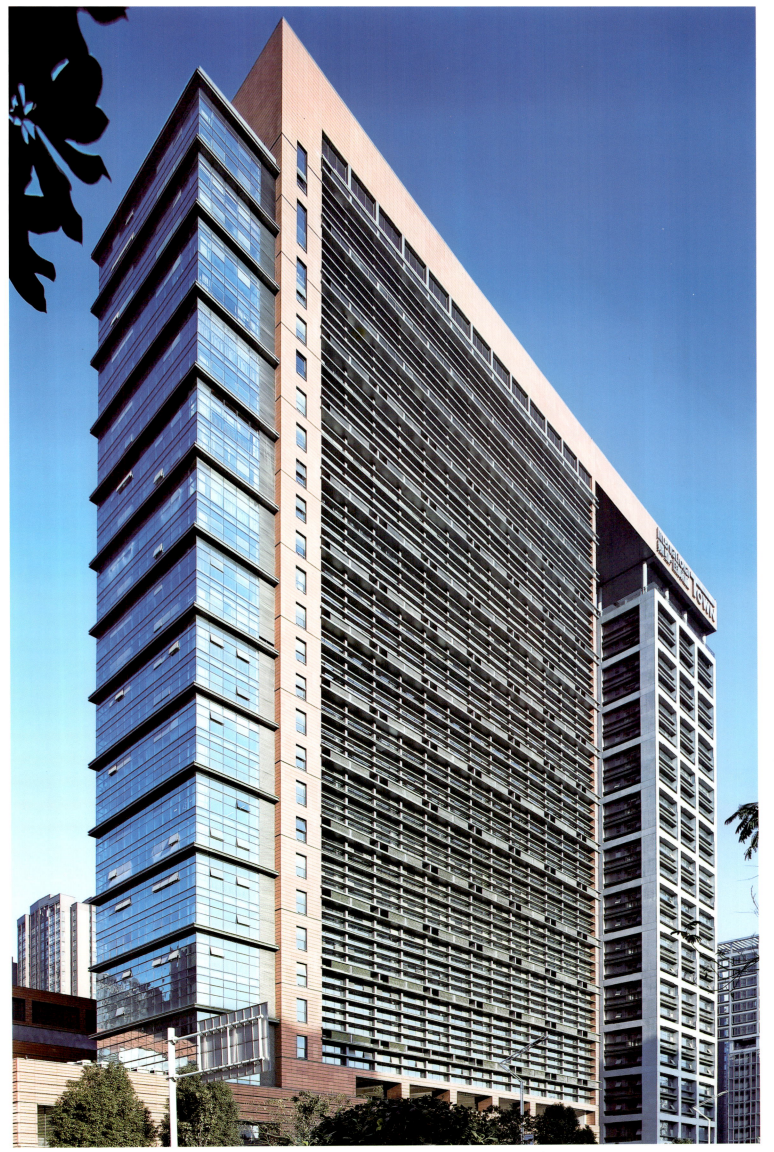

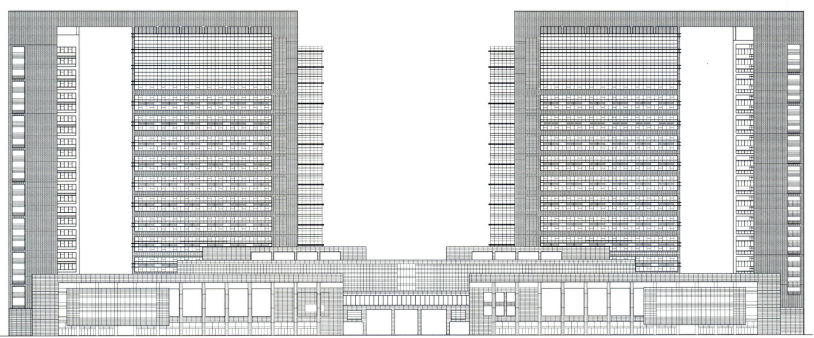

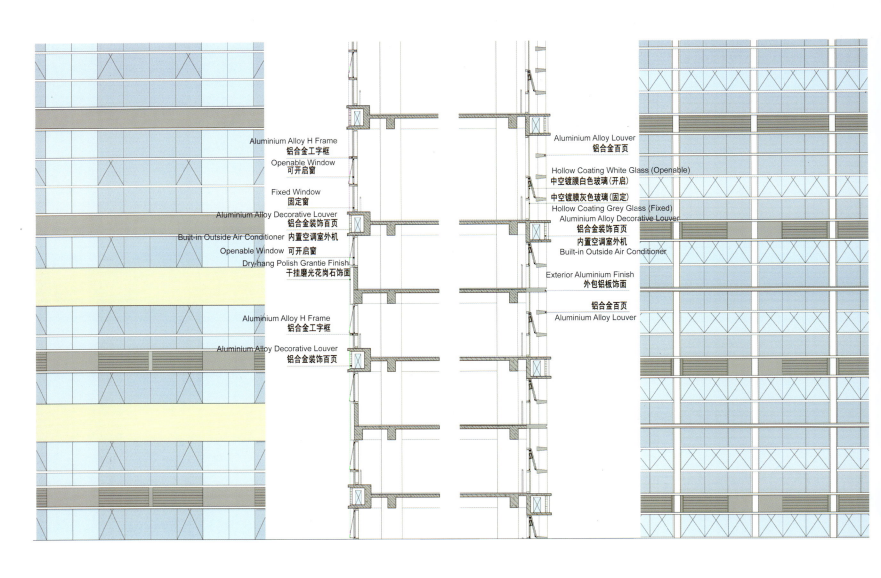

Aluminium Alloy H Frame
铝合金工字框

Openable Window
可开启窗

Fixed Window
固定窗

Aluminium Alloy Decorative Louver
铝合金装饰百页

Built-in Outside Air Conditioner 内置空调室外机

Openable Window 可开启窗

Dry-hang Polish Grantie Finish
干挂磨光花岗石饰面

Aluminium Alloy H Frame
铝合金工字框

Aluminium Alloy Decorative Louver
铝合金装饰百页

Aluminium Alloy Louver
铝合金百页

Hollow Coating White Glass (Openable)
中空镀膜白色玻璃(开启)

中空镀膜灰色玻璃(固定)
Hollow Coating Grey Glass (Fixed)
Aluminium Alloy Decorative Louver

铝合金装饰百页
内置空调室外机
Built-in Outside Air Conditioner

Exterior Aluminium Finish
外包铝板饰面

铝合金百页
Aluminium Alloy Louver

北立面
NORTH ELEVATION

剖面
SECTION

南立面
SOUTH ELEVATION

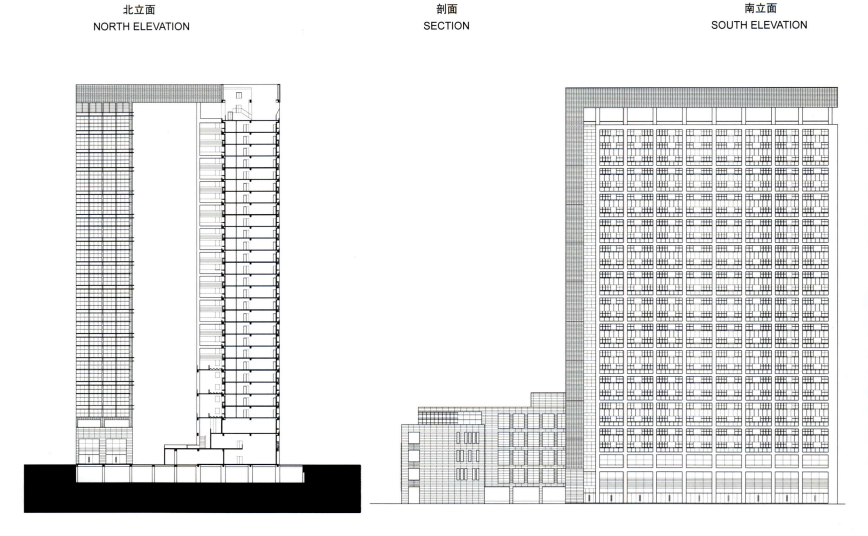

该设计首先注重良好的功能布局和交通组织，强调步行空间的塑造，将机动车流线设置在规定范围内。机动车主要由辅街进入，庭院铺地的变化和路型的设计限定了机动车的流线，从而将车行和步行区域区分开来，保证了交通的通畅，方便贵宾直达入口，避免了内部空间完全被机动交通所破坏。该项目以步行交通为主，并以人车混行的方式组织交通，提升了环境的品质。

内部环境的处理注重生态环境品质。圆通广场采用短进深办公平面，便于直接通风和采光，同时设计师还对自然光的利用进行了研究。酒店式办公楼采用近似对称的中廊式布局，南北两个立面均结合空调室外机的放置，设计了若干水平构件，但其分布方式却因内部空间对阳光、天光的不同需要而大相径庭。南立面上为遮挡夏季强烈的阳光直射，尤其是近午时的光线，放置空调室外机的水平构件均设于室内空间的上部，同时利用偶数层的楼板出挑，并结合其他百叶式水平构件，进一步避免阳光直射。

与之对应的北立面则强调对天光的利用，放置空调室外机的水平构件设于室内空间的下部，形成的宽大窗台与其他水平构件一起很好地起到折射、漫反射天光的作用，使北向的室内空间获得充沛、均匀的自然光。另外，包裹建筑物的外立面也南北有别：南立面退到水平构件以内，以减弱阳光对室内的直晒；北立面凸出，基本露在水平构件之外，为的是多利用自然天光。

建筑外立面选用红色陶土板、清水混凝土与玻璃幕墙等绿色环保材料，传统质朴的陶土板与现代明快的玻璃相互辉映，仿佛古老与现代的对话，自然和科技的共生，为整个中心商务区建设注入了文化与生态内涵。陶土板幕墙主要采用210mm×1530mm 和305mm×315mm 两种规格的陶土板及相配套的陶土百叶，将大尺度的建筑体量与精美的肌理划分结合起来，并对更细致的工艺设计和节点进行重点推敲。

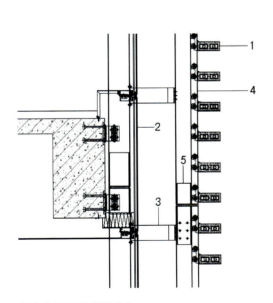

铝合金百叶垂直节点 1
Aluminium Alloy Louver Vertical Joint 1

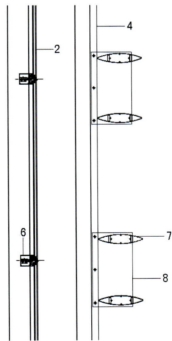

铝合金百叶垂直节点 2
Aluminium Alloy Louver Vertical Joint 2

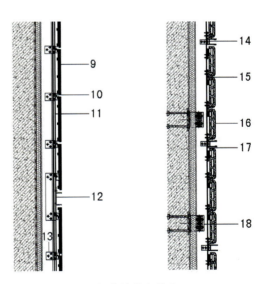

陶土板幕墙纵向节点
Terracotta Panels Parallel Joint

1. 150mm × 60mm × 330mm 矩形陶土百叶片
   150mm×60mm×330mm Rectangle Clay Louver Blade

2. 6mm+12A+6mm 钢化中空镀膜玻璃
   6mm+12A+6mm Tempered Hollow Coated Glass

3. "T" 形铝合金型材
   T-shaped Aluminium Alloy Material

4. 铝合金盖板
   Aluminium Alloy Cover Plate

5. 铝合金芯套
   Aluminium Alloy Cover Retainer Plate

6. 铝合金横梁
   Aluminium Alloy Beam

7. 铝合金百叶片
   Aluminium Alloy Louver

8. 5mm 厚钢板
   5mm Steel Plate

9. 14mm 厚进口陶土板
   14mm Imported Terracotta Panels

10. 铝合金横梁
    Aluminium Alloy Beam

11. 60mm × 80mm × 4mm 镀锌钢通
    60mm×80mm×4mm Galvanized Steel

12. L 形铝合金组装为槽形装饰线条
    L-shaped Aluminium Alloy Groove Shape Decorative Stripe

13. 50mm × 70mm × 4mm 铝角码
    50mm×70mm×4mm Aluminium Connector

14. 槽形铝合金装饰线条
    Aluminium Alloy Groove Shape Decorative Stripe

15. 不锈钢陶土板挂码
    Stainless Steel Terracotta Panels Weights

16. 40mm 厚进口陶土板
    40mm Imported Terracotta Panels

17. 60mm × 80mm × 4mm 镀锌钢通
    60mm×80mm×4mm Galvanized Steel

18. 60mm × 100mm × 7mm 镀锌钢角码
    60mm×100mm×7mm Galvanized Steel Connector

ACADEMY OF CHINA WANDA

河北廊坊万达学院

CHINA GREEN BUILDING 3 STAR CERTIFIED 中国绿色建筑三星级认证

ARCHITECT
HYHW Architecture Consulting LTD

PROJECT TEAM
HYHW, Ove Arup, Hassell, China BCEL

LOCATION
Langfang, Hebei Province, China

AREA
133,300 m²

PHOTOGRAPHER
Shu He Photography

SUSTAINABLE & GREEN FEATURES 绿色特征

- *Use of ventilation and heat recovery system* ／ 使用排风热回收系统
- *Highly efficient water system and recyclable materials* ／ 高效节能的水系统和可回收材料利用
- *Large use of natural ventilation and daylight* ／ 充分使用自然通风和自然光

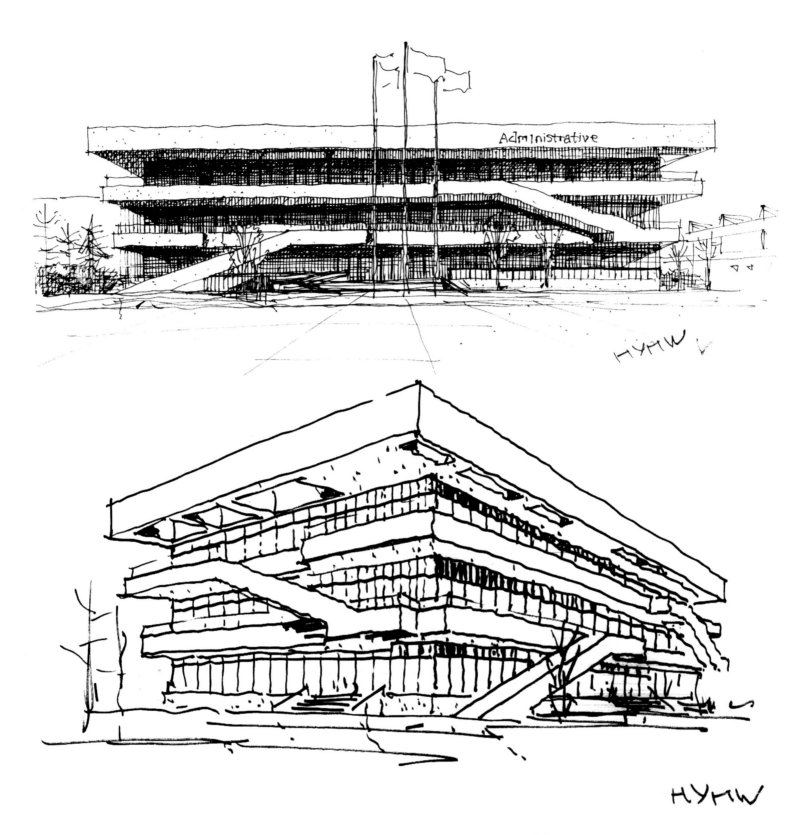

Wanda Academy is a higher education facility for full time occupational studies, operated by Wanda Group for employee training. It aims at growing into an intermediate and advanced personnel training base with a variety of functions. Covering an area of about 133,333 square meters, the campus is located in the northeast of Langfang, with Liyuan Road to the east and Garden Road to the north. Wanda Academy has an overall land area of 133,300 square meters, and the total GFA of 128,100 square meters. The Wanda Academy is being constructed in two phases. The total GFA of phrase one is 52800 square meters, including an over ground area of 49,300 square meters. The project includes a main teaching building, a library, an administration building, a gymnasium, accommodation, canteens, exhibition halls, an information center, walls near the gate and reception room, and stadium and audience area.

The whole architectural design symbolizes Wanda Group's corporate motto, "Efficiency and Rigor". Stone is used as the primary material on the façade to create a solid, simple, clean and modern aesthetic.

The main teaching building emphasizes vertical structure, giving the building a positive and academic atmosphere. It forms a circle with the lecture hall and the multifunction building on either side to its south, and becomes the most important feature of the Wanda Academy.

The administration building enjoys extended spaces and large open balconies. It is a focal point of the square on the south side. The proportions of the square and horizontal elements versus the height were the main consideration when refining the building design.

The exhibition building is cube in shape, with the entrance as the focus of visual and psychological attention. The LED lighting system adorns the elevation which creates a peaceful and romantic environment at night.

The canteen, together with the sunken courtyard on its east is an important outdoor public space in the plan. Advanced Sports facilities including a swimming pool and fitness gym demonstrate sculpture-like power and strength.

The Data Center is the only building using metal on the façade to show a digitalized feature. A linear-shaped space is formed by student accommodation, teaching building, canteen and data center. This functional "Street" space will be the main social public space of the college.

HYMW

万达学院是万达集团为培训员工而成立的全日制高等职业院校。将成为万达集团一个多层次、多功能的高中级人才培养和培训基地。校园位于廊坊市东北部，占地面积约 133333 平方米，东临梨园路，北临花园道。万达学院总用地面积约为 13.33 万平方米，总建筑面积约 12.81 万平方米。学校按两期分期建设，一期总建筑面积为 5.28 万平方米，其中地上建筑面积为 4.93 万平方米。建筑项目包括教学楼、图书馆、行政办公楼、体育馆、宿舍、食堂、规划展览及酒店样板展示馆、信息中心、大门围墙及传达室、体育运动场地及看台。

整体建筑设计体现了万达集团"高效严谨"的企业特质。主建筑立面均以石材为主，表现坚实沉稳、简洁现代的设计感。

主体教学楼设计强调竖向构成，挺拔向上而又富有书卷气息。向南与两侧的报告厅和多功能教室围合成庭院，成为学院最重要的建筑元素。

行政楼造型重点表现向上收放的水平阳台，表达舒展和开放的气质，是其南侧广场的构图焦点，其强调水平的特质配合了广场的尺度。

展览建筑是正方形几何体，入口庭院空间给人强烈的视觉和心理震撼。其四面外墙上镶嵌的 LED 点灯，在夜晚将赋予学院浪漫、宁静的气息。

餐厅建筑与其东侧的下沉庭院相互呼应，共同成为规划设计中重要的室外休闲公共空间。体育建筑包括了游泳、健身等先进的体育设施，建筑形态强调有力的雕塑化特点。

数据中心是众多建筑中唯一以金属作为主要外装材料的建筑，用以表现数字化的个性。值得一提的是，由学员宿舍、教学楼、餐厅和数据中心形成的线性空间，呈现出"街道"的空间和功能原型，也将成为学院重要的交往性公共空间。

LONGYUE RESIDENCE Ⅲ

龙悦居三期

CHINA GREEN BUILDING 2 STAR CERTIFIED 中国绿色建筑二星级认证

ARCHITECT
Capol

LOCATION
Shenzhen, Guangdong Province, China

SUSTAINABLE & GREEN FEATURES 绿色特征

- *Use of solar hot water system and solar lamps* / 使用太阳能热水系统和太阳能灯具
- *Use of rainwater collecting and recovery system and grey water recycling system* / 使用雨水收集利用系统和中水处理回收系统

This project is one of the ten people's livelihood projects for the 30th anniversary of Shenzhen Special Economic Zone. Designed and built in industrialized residence system, this project aims to create a residential demonstration community and a housing industrialization demonstration project in Shenzhen, which hopefully will motivate more real estate development enterprises to use new technologies to save energy, land, water power, raw materials and therefore to protect the environment, channeling Shenzhen Housing Industry into an intensive, resource-saving and ecological development mode.

Six high-rise residences placed around the site and a central courtyard form the layout. Reserved mountain and landscape are connected together through the axis of the city's traffic network. And the design takes full advantage of the original mountain, transforming it into a park for visitors. Designers combine it with public landscape to build a beautiful, harmonious and well-structured community.

This project is built in an industrialized housing system, meeting the standards of industrial construction in component manufacture, transportation, site operation and component standardization. The concept

of industrial modular production runs through the design, which tries to reduce the species of prefabricated parts and molds, improves the molds' utilization rate and reduces the cost of production under the principle of "modularization of apartment design, standardization of typical floor design and large scale overall planning". Prefabricated parts are made from steel template in factory, which reduces the use of wooden template in site operation so that it can save natural resources. Production cycle would be greatly shortened by the use of the more efficient site-assembled parts. The quality of decoration can be guaranteed through integrating indoor design and large-scale centralized purchasing, together with safer and more environmental friendly decoration materials. Therefore, we can avoid wasting materials in re-decoration and save materials to the greatest extent.

Green architecture design combines simulation analysis methods using perfect green technologies such as solar energy hot-water system, constructed wetland and rainwater collection, which meets the standard of one star national green architecture, Shenzhen green architecture platinum standard and also to satisfy the needs for minimizing the use of energy, materials, land and water.

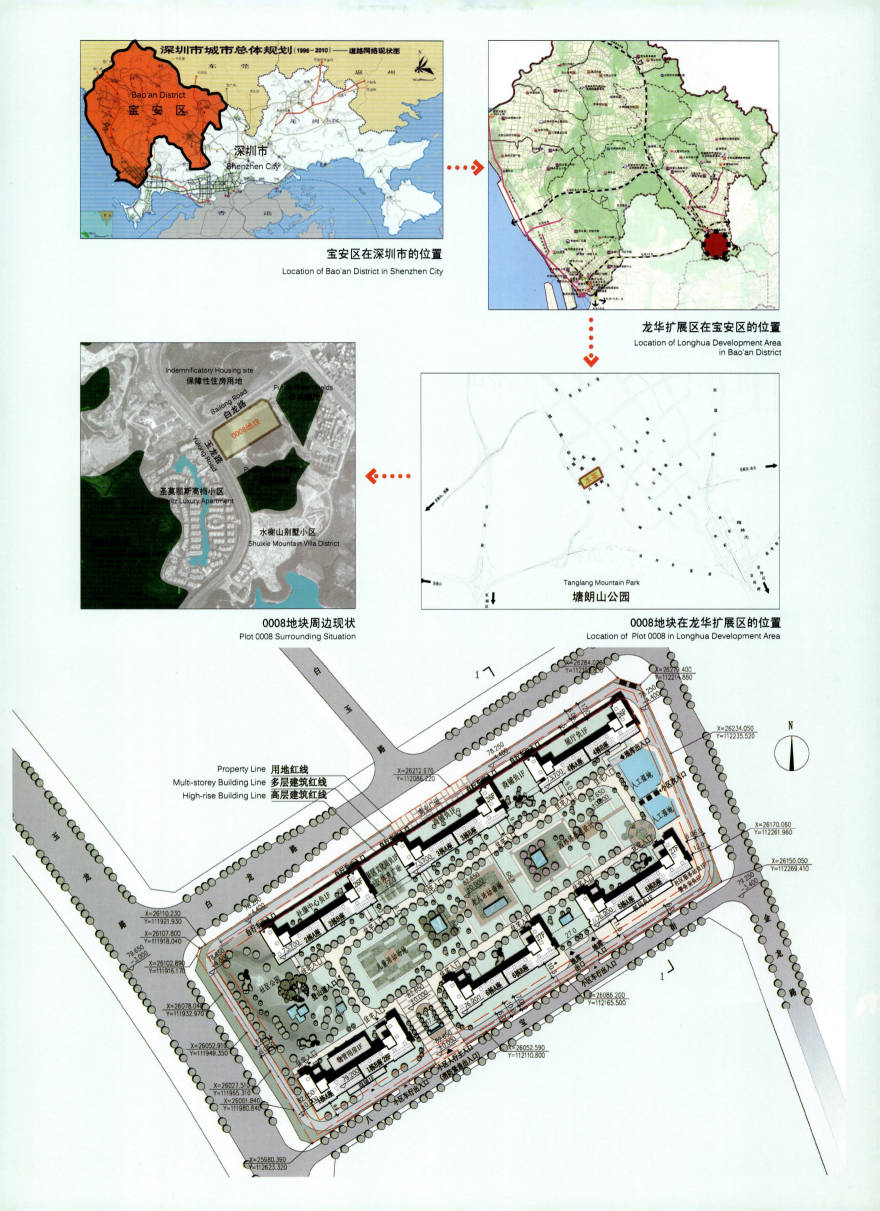

宝安区在深圳市的位置
Location of Bao'an District in Shenzhen City

龙华扩展区在宝安区的位置
Location of Longhua Development Area
in Bao'an District

0008地块周边现状
Plot 0008 Surrounding Situation

0008地块在龙华扩展区的位置
Location of Plot 0008 in Longhua Development Area

Property Line 用地红线
Multi-storey Building Line 多层建筑红线
High-rise Building Line 高层建筑红线

Energg-efficient & Eco-friendly
节能环保

建筑外墙采用国产中档节能外墙涂料和高效保湿材料。

采用铝合金门窗，临主要干道的卧室窗户采用双层中空玻璃。

屋面采用保温屋面技术，部分屋面采用种植屋面。

Domestic mid-range energy-saving coating and highly-efficient thermal insulation materials are used on the façade.

Aluminium alloy doors and windows are used. Bedrooms close to the main passage use double-layer hollow glass on the windows.

Heat preservation Technology is used on the roof. Some are planted roof.

Full Use of Small Space
小空间合理利用

小空间，大学问。小空间的合理利用，在保障型住房设计中尤为重要，需要充分合理地利用每一寸空间。设计中要考虑人体工学，体现舒适性。

Small space, great knowledge. Full use of small space is very important in the design of indemnificatory housing. Every inch of space should be fully used. Ergonomic is considered in the design to ensure comfort.

One-time Refined Decoration
全部精装修

装修一步到位，避免二次装修的资源浪费。严格控制施工质量，确保装完不修。

Decoration is done at one step to avoid resource waste by second decotation. Constrution quality is strictly controlled to ensure one-step decoration.

Public Space
公共空间

架空层、入户大堂等公共空间的作用被充分挖掘，通过精心的设计，让每一个空间都能成为最佳的休闲娱乐场所，为回家的你筑就理想之路。

Empty space, lobby and other public space are fully made use of. Meticulous design makes every space become best places of leisure and pave the way for you to go home.

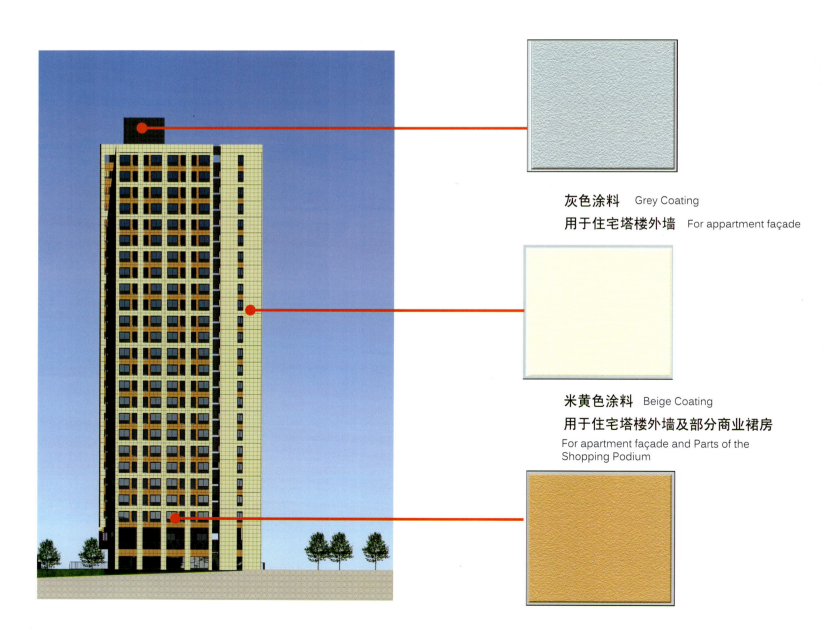

灰色涂料　Grey Coating
用于住宅塔楼外墙　For appartment façade

米黄色涂料　Beige Coating
用于住宅塔楼外墙及部分商业裙房
For apartment façade and Parts of the Shopping Podium

仿木色涂料　Wood Coating
用于住宅塔楼外墙　For appartment façade

本项目 PC（预制混凝土）外墙在保证立面美观、大方的前提下，立面开洞（阳台及窗）尺寸大小统一，外墙选择光整预制面，避免了因形体及立面的凹凸带来的工业化构件复杂，外饰面采用建筑涂料后施工法，有效地减少了构件生产模具数量，降低了预制阶段对工艺和时间的要求，实现了降低成本的目的。

立面面层采用涂料，基于 PC（预制混凝土）外墙的平整性，可最好地发挥涂料的效果，避免因常规外墙基层开裂带来的效果上的破坏；与面砖及石材面层相比，降低了工艺及施工的复杂性，防止施工过程中不确定因素对外墙饰面造成的污染，从而降低了成本；此外，涂料在立面形体及色彩的划分上更为灵活。

本项目外廊及栏杆亦按照采用标准化生产、现场装配的方式进行设计与施工。

In the premise to make sure the precast concrete façade will look beautiful and elegant, the façade punches (balconies and windows) have the same size. The façade uses finished prefabricated surface, which avoids industrialized structural complexity caused by the concave-convex of shape and façade. Outside finish uses the construction method of after coating, which effectively reduces the number of component production die and the demands of technology and time in the period of prefabrication, and then reduces the cost.

The façade surface uses coatings. On the basis of the smooth precast concrete façade, the effect of coating is evaluated and this avoids damages by the effect of of normal façade base crack. Compared with façade of face brick and stone, coating façade reduces the complexity of technology and construction and prevents the pollution on the façade caused in the construction and reduces the cost. Besides, coating is more flexible in distinguishing shapes and colors.

The verandas and handrails also adapt standardized production, design and construction in the way of on-site assembly.

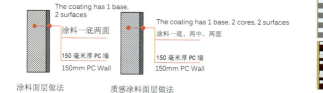

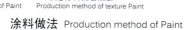

涂料做法 Production method of Paint

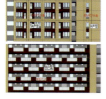

立面开洞统一 The façade punches have the same size

工业化立面效果 Effect of industrialized façade

Use of materials and resources	Use of energy	Use of water	Operations management
材料资源利用	能源利用	水资源利用	运营管理

建筑材料中有害物质含量符合国家标准和《建筑材料放射性核素限量》的要求。不使用国家及深圳市建设行政主管部门公布的限制、禁止使用的建筑材料及制品。

建筑造型要素简约，无大量装饰性构件。

将建筑施工过程中产生的固体废弃物分类处理和回收利用，回收利用率不低于30%。

现浇混凝土采用预拌混凝土。

充分利用自然条件，整个规划通风采光良好，节约采光及空调用电。采用分散式空调器，能效比符合《深圳市居住建筑节能设计标准实施细则》的规定。

制定水系统设计方案，统筹、综合利用各种水资源。地区水资源状况、气象资料、市政设施情况的说明；用水定额的确定、用水量估算及水量平衡表的编制；非传统水源利用方案；采用节水器具、设备和系统的方案。

制定并实施节能、节水、节材与绿化管理制度。为所有住户提供环境维护指导手册，定期进行培训与宣传活动。

居住建筑水、电、燃气分户、分类计量与收费。

制定垃圾管理制度，对垃圾物流进行有效控制，对废品进行分类收集，防止垃圾无序倾倒和二次污染。

设置密闭的垃圾容器，并有严格的保护清洗措施，生活垃圾采用袋装化存放。

The content of harmful substances in the building materials meets the national standard and the demands of the "Limits of Radion Nuclide in Building Materials". Building materials and products which are limited or forbidden by the national and Shenzhen construction administrative departments are not used.

The building appearance is simple, without plenty of decorative components.

The solid wastes produced in the construction period are sorted out and recycled. The recovery rate is not under 30%.

Cast-in-situs concrete uses the premixed concrete.

The natural condition is made full use of. The project is with good natural ventilation and natural lighting, which reduces consumption by artificial lighting and air contiditioning. Distibuted air conditioners are used, which meets the requirements of "Detailed Rules for Application in Shenzhen for Residential Buildings Energy Efficiency Design Standard".

Plans for water system design are made to take good advantage of kinds of water resources: district water resource status, meteorological data, service condition of public facilities; determination of water quota, water quota estimation and water quota balance sheet compilation; untradtional water resources plan; and plans to use water-saving tools, equipments and systems.

Systems of energy saving, water saving, material saving and landscape management are developed and implemented. Environmental protection guide books are offered to all residents and campaigns are held regularly.

Water, electricity and gas are measured and charged by household and categories.

Waste management system is made to effectively control waste transportation and classify and collect waste, in order to avoid disorder dumping and secondary pollution.

Airtight waste containers and strict cleaning measures are used. Household waste is collected in bags.

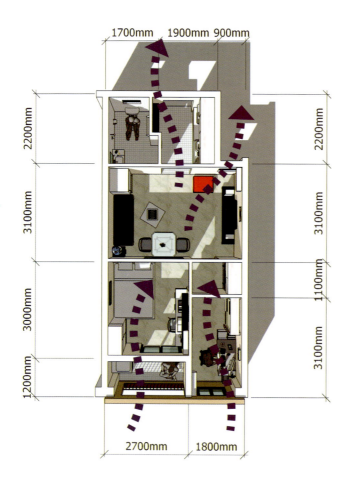

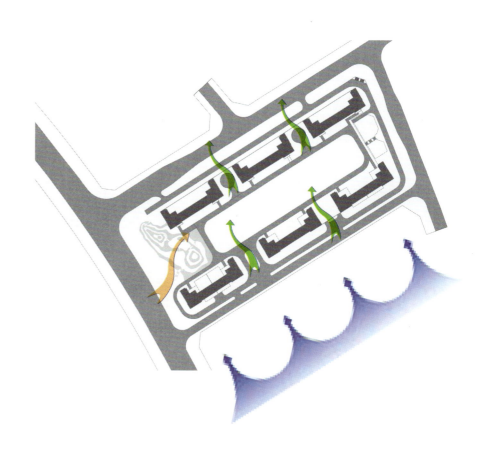

"可呼吸式"建筑单体设计，合理组织户内通风。户型设计上，充分考虑客厅、内天井、阳台的对位布置关系，合理设置门、窗户的开启位置，组织南北通透的气流。

Breathing architecture design reasonably controls the indoor ventilation. The house type design takes a full consideration on counterpoint decoration connections among living room, indoor courtyard and balcony. The open position of doors and windows are reasonably set to organize north-south ventilation.

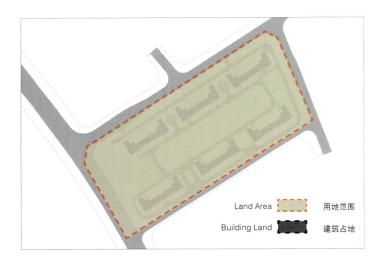

点式高层的规划理念，有效节约建筑占地

Point high-rise masterplan design efficiently saves building land

Land Area 用地范围
Building Land 建筑占地

外挂预制 PC 外墙系统，工厂化、标准化、产业化

Plug-in prefabricated PC façade system is industrialized, standardized and commercialized

菜单式装修，减少二次装修的浪费

Menu-like decoration reduces waste of second decoration

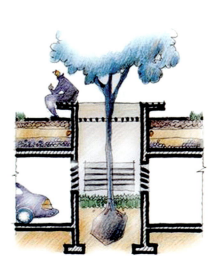

　　半地下室顶板上局部开设洞口，将自然光及通风引入半地下停车空间，并可在开口部位的地下室底板上种植绿化，形成立体的多层次生态车库，沿建筑外墙的半地下室上部设玻璃百叶，将光线和风从侧面引入地下车库，与庭院中开口处形成空气对流。

　　There are entrances on the top-plates of the semi-basement, in order to bring in the natural daylight and ventilation to the semi-basement. Plants are also used on the bottom plates of the basement which have entrances, in order to make a steric and multi-level ecological parking garage. Upper plates of the semi-basement which is close to the building façade are installed with glass louver, in order to bring in daylight and ventilation to underground garage and make cross-ventilation with the courtyard entrances.

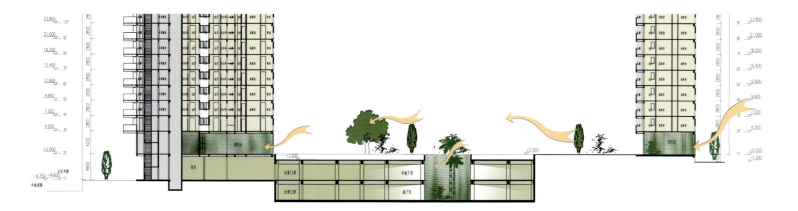

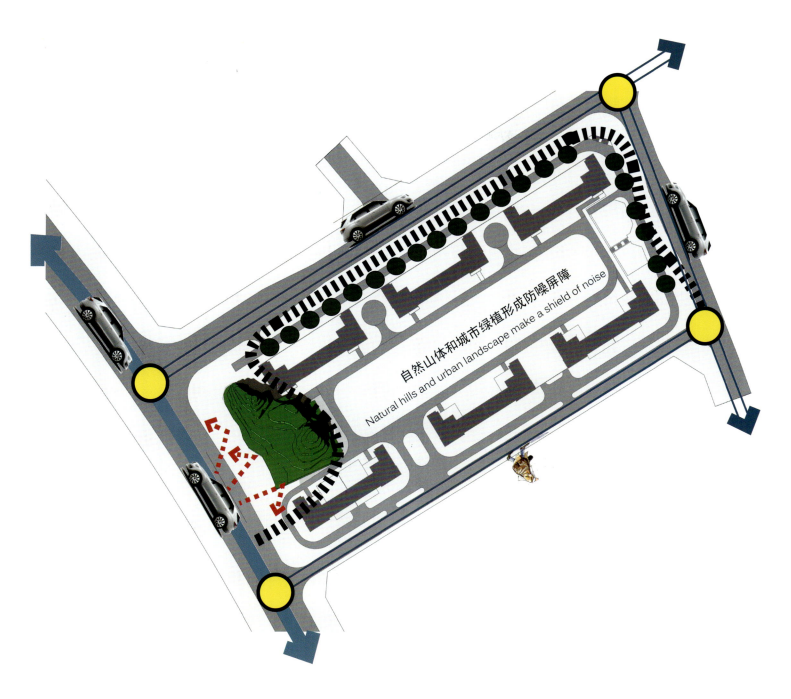

自然山体和城市绿植形成防噪屏障
Natural hills and urban landscape make a shield of noise

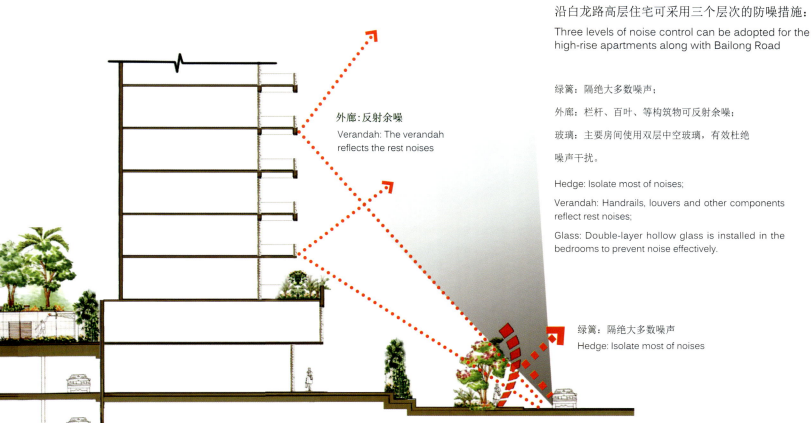

外廊：反射余噪
Verandah: The verandah reflects the rest noises

绿篱：隔绝大多数噪声
Hedge: Isolate most of noises

沿白龙路高层住宅可采用三个层次的防噪措施：
Three levels of noise control can be adopted for the high-rise apartments along with Bailong Road

绿篱：隔绝大多数噪声；

外廊：栏杆、百叶、等构筑物可反射余噪；

玻璃：主要房间使用双层中空玻璃，有效杜绝

噪声干扰。

Hedge: Isolate most of noises;

Verandah: Handrails, louvers and other components reflect rest noises;

Glass: Double-layer hollow glass is installed in the bedrooms to prevent noise effectively.

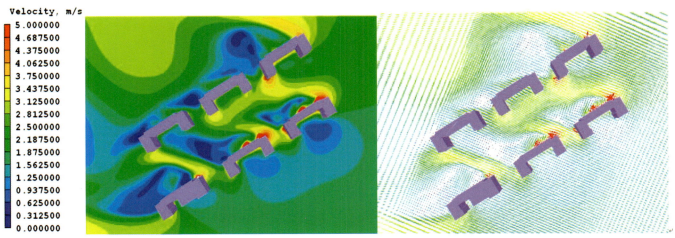

人行位置（1.5m）风速图　Pedestrian location (1.5m) Wind Speed Chart

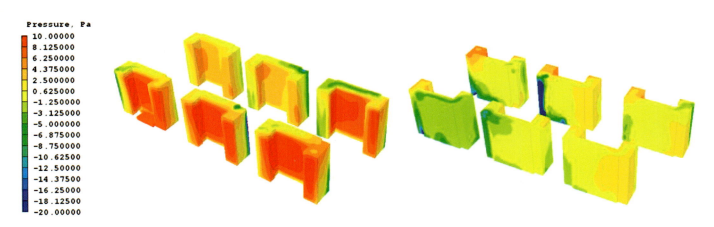

建筑迎背风面表面压力分布图　Scattergram of windward side and leeward side pressure

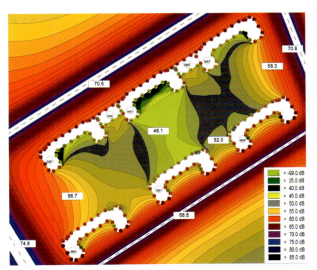

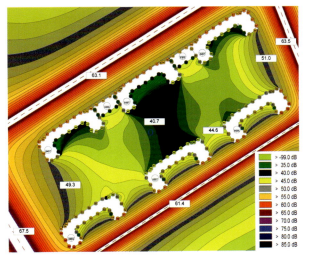

小区人员活动高度处昼间环境噪声（dB（A））分布图　　　　小区人员活动高度处夜间环境噪声（dB（A））分布图

Scattergram of noise (dB(A)) distribution of resident activity places in day/night

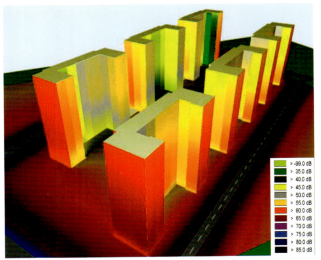

东南向建筑表面声压级（dB（A））分布图（昼间/夜间）

Scattergram of sound pressure level (dB(A)) of southeast building façade (day/night)

该项目是深圳经济特区建立 30 周年十大民生项目之一，应用工业化住宅体系设计建造，旨在创建深圳市建筑示范小区和住宅产业化示范项目，以带动更多开发企业积极采用"节能、节地、节水、节材、环保"新技术，促进深圳市住宅产业向集约型、节约型、生态型发展模式转变。

规划设计以六栋高层住宅沿地块周边布置，结合中心庭院形成围合式布局。一条顺应城市路网的空间轴线将地块内的保留山体和园林景观串联在一起，并有效利用原始山体，将其改造成为登山公园，与公共园林景观组建优美的富有层次感的和谐社区。

本项目应用工业化住宅体系设计建造，满足了工业化生产方式中构件生产、运输、现场施工和部品标准化等要求。在进行建筑设计时，由始至终贯穿着工业化生产方式

的模块化设计理念，坚持"户型单体设计模块化、标准层平面设计标准化、总体规划布局规模化"的原则，尽量减少预制构件和模具的种类，大大提高模具的利用率，降低生产成本。预制构件在工厂采用钢模板生成，减少现场施工中木模板的使用，进而节约了自然资源。现场装配化施工效率更高，建造周期大幅缩短。一体化室内精装设计施工，大规模集中采购，装修材料更安全、环保，标准化的装修保障了装修质量。避免了二次装修对材料的浪费，最大程度节约材料。

绿色建筑设计结合模拟分析的方式，应用太阳能热水系统、人工湿地、雨水收集等完善的绿色技术，达到了国家绿色建筑一星级、深圳市绿色建筑铂金级的标准，最大程度满足节能、节材、节地、节水等方面的要求。

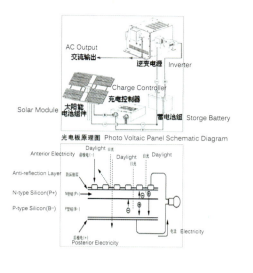

AC Output
交流输出 Inverter
Charge Controller
充电控制器
Solar Module
太阳能电池组件
雷电池组 Storage Battery
光电板原理图 Photo Voltaic Panel Schematic Diagram

Anterior Electricity
Daylight 日光 Daylight Daylight
Anti-reflection Layer 减反射膜
N-type Silicon(P+) N型硅(磷+)
P-type Silicon(B−) P型硅(硼−)
电流 Electricity
Posterior Electricity

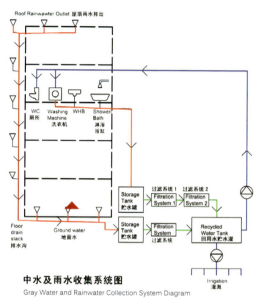

Roof Rainwawter Outlet 屋顶雨水排出

WC 厕所　Washing Machine 洗衣机　WHB　Shower Bath 淋浴 浴缸

Floor drain stack 排水沟　Ground water 地面水

Storage Tank 贮水缸 → 过滤系统 1 Filtration System 1 → 过滤系统 2 Filtration System 2

Storage Tank 贮水缸 → 过滤系统 Filtration System → 回用水贮水缸 Recycled Water Tank

Irrigation 灌溉

中水及雨水收集系统图
Gray Water and Rainwater Collection System Diagram

电气设计　Electical Design

1. 生活水泵采用变频控制；
Water pumps are under frequency conversion control.

2. 小区路灯照明、环境照明等考虑采用内置智能光控与时控装置的自带太阳能电池板的太阳能灯具。灯具可全自动开关，无需人工操作看守；
Street lights and ambiant lighting use built-in solar lamps with solar panels which are intelligently light controlled and time controlled. All have automatic switch, so people guard is not needed.

3. 小区半地下室结合建筑和环境尽量采用自然采光，减少人工照明，同时按白天／夜晚场景细分控制回路以达到节能的目的；
Combining with the building and the environment, the semi-basement takes full advantage of daylight and reduces artificial lighting. Control loop is subdivied by day and night to reduce energy consumption.

4. 住宅走道、楼梯间采用节能自熄开关；
Enery-saving self-extinguishing switches are installed in residential corridor and stairs.

5. 照明、水泵、动力等部分能耗实现分项和分区域计量。
Lighting, water pump and power consumption are measured by categories and regions.

暖通设计　Heating and Ventilation Design

1. 住宅塔楼防烟楼梯间及其前室利用敞廊和可开启外窗自然通风排烟，可以提高建筑实用面积，节省造价和耗电，维护管理方便；
Smoke prevention stairs and prechambers of the residential towers use loggia and openable windows for natural ventilation and smoke extraction. This increases building usable area, saves cost and electricity consumption and maintains easy management.

2. 卫生间设置风量可调的排风装置，通风排烟系统设有防倒灌措施；
Air exhaust system with adjustable air volume is installed in the bathroom. The ventilation and smoke extraction systems have no backflow prevention measures.

3. 通风换气系统设计避免气流短路，新风由洁净区域流向较为不洁的区域。在通风口安装防虫网；
Ventilation and air change systems avoid air short circuit. The new air flows from clean areas to unclean areas and insect proof nets are installed on the vents.

4. 通风排烟系统风机单位风量耗功率满足节能规范要求，并选用高效、节能、低噪声的风机；
Fans consumption in ventilation and air change systems meets the national requirements and standards and all fans are highly efficient, energy saving and low-noise.

5. 分体式房间空调器能效值达到能效等级 2 级要求；
Energy efficiency grades of Split-type air conditioner in the rooms is ranked Level 2.

6. 空调室内机安装位置有利于房间气流均匀；
The position of indoor air conditioners is good for even air flow.

(1) 统一设计分体式房间空调器的安放位置和搁板构造；
The indoor split-type air conditioner installation positions and shelf structures are designed uniformly.

(2) 避免多台相邻室外机吹出气流相互干扰，避免高层建筑烟囱效应对高区室外机换热效率的影响；
Avoid the blowout air flows from many neighboring outdoor air conditioners affecting each other and avoid chimney effect of high-rise building on heat exchange efficiency.

(3) 室外机安装位置避免对相邻住户造成热污染和噪声污染；
Outdoor air conditioner positions avoid producing heat pollution and noise pollution to neighboring residents.

7. 空调设备凝结水有组织排放和收集；
Condensed water from air-conditioners is organized to blowoff and collect.

8. 发电机房烟气、厨房排油烟等经过净化处理后排放或高空排放。
Smokes from generator room and kitchen are let out after purification treatment.

运营管理　Operations management

1. 制定并实施节能、节水、节材与绿化管理制度。为所有住户提供环境维护指导手册，定期进行培训与宣传活动；
Systems of energy saving, water saving, material saving and landscape management are developed and implemented. Environmental protection guide books are offered to all residents and campaigns are held regularly.

2. 居住建筑水、电、燃气分户、分类计量与收费；
Water, electricity and gas are measured and charged by household and categories.

3. 制定垃圾管理制度，对垃圾物流进行有效控制，对废品进行分类收集，防止垃圾无序倾倒和二次污染；
Waste management system is made to effectively control waste transportation and classify and collect waste, in order to avoid disorder dumping and secondary pollution.

4. 设置密闭的垃圾容器，并有严格的保洁清洗措施，生活垃圾采用袋装化存放。
Airtight waste containers and strict cleaning measures are used. Household waste is collected in bags.

EASTERN HARBOR
INTERNATIONAL TOWER

东方海港国际大厦

ARCHITECT
Jeffery Heller

FIRM
Heller Manus Architects

LOCATION
Shanghai ,China

SUSTAINABLE & GREEN FEATURES　绿色特征

- *Large use of natural ventilation and daylight* ／ 大量使用自然通风和自然光
- *Use of heating ventilation air conditioning* ／ 使用暖通空调系统
- *Large use of water-saving systems* ／ 大量使用节水装置
- *Green roof to recycle rainwater* ／ 绿色屋顶用于雨水收集

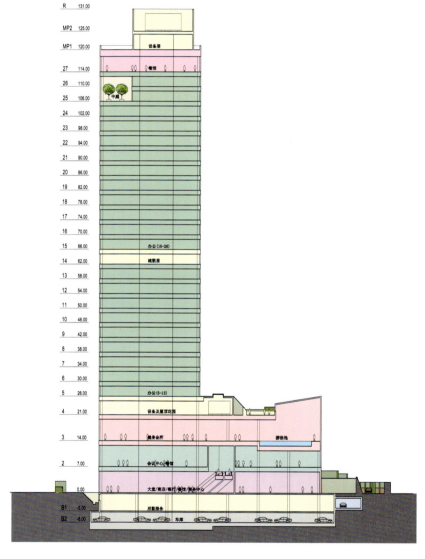

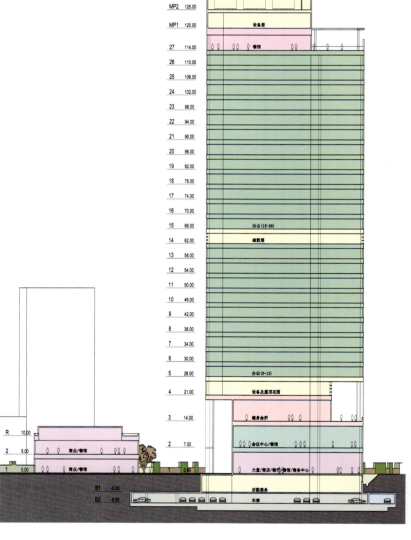

Eastern Harbor International Tower is a multi-use building consisting of a 120-meter high office tower with rooftop garden over a 21-meter high podium containing a health club with pool, conference and banquet facilities and ground floor retail as well as a separate 10 meter high retail building. A restaurant and terrace are located on the top floor of the tower allowing for commanding views of the skyline. The design incorporates an indoor environment with essential sustainable features such as energy and water conserving measures, recycling, and renewable materials that enabled it to become the first LEED Gold certified high-rise in Shanghai, China.

It was the objective of the owner and the design team to create a functional, advanced landmark building with energy efficiency and environmental protection features while maintaining a level of fire

东大名路
East Daming Road

双向机动车流	Bidirectional Vehicle flow
单向机动车流	Unidirectional Vehicle flow
消防车环道	Fire-fighting truck Circuit
客运电梯	Passenger elevator
服务兼消防电梯	Service and fire elevator
疏散楼梯	Evacuation stairs

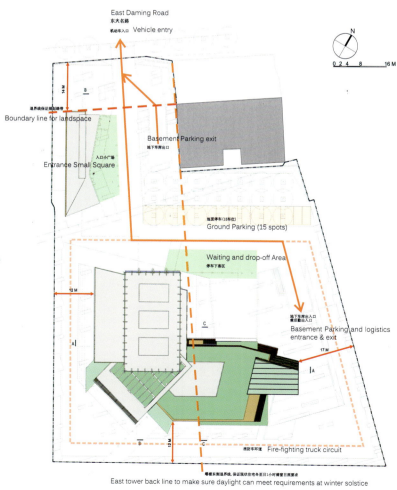

East Daming Road
东大名路
机动车入口 Vehicle entry

Boundary line for landspace

Entrance Small Square

Basement Parking exit

Ground Parking (15 spots)

Waiting and drop-off Area
停车下客区

Basement Parking and logistics
entrance & exit

Fire-fighting truck circuit

East tower back line to make sure daylight can meet requirements at winter solstice

protection and security consistent with the National Standards of the People's Republic of China.

• HVAC design: complex ground-source heat pump system , efficient screw chiller units, heat recovery make-up air units, VAV system (pressure control), water pump inverter (cooling water, chilled water, ground source measuring pump), cool tower double speed fan;

• CO_2 sensor control system(ventilation on demand);

• Nature ventilation in excessive season, taking cold for free;

• Switch-type sensor compensated by sunlight outside the external windows of the construct;

• Optimizing LPD with efficient energy-saving lights;

• Making full use of natural light, sound light operated delay switch in pavement. According to the natural lighting in every room, the illumination is controlled in different zone and group;

• Personnel sensors;

• Efficient low-energy transformer, reactive compensation measure on the spot to provide power factor;

• Choosing high-performance glass, adjustable windows in every floor to natural ventilate and to reduce the usage time of air condition;

• External building envelope, glass curtain wall made by LOW-E hollow toughened glass. Stone curtain wall, 50mm YTong light concrete slab, the unit curtain wall system of external thermal insulation integration;

• Using the suitable shading form, effectively control the glare. Glass curtain wall with horizontal shutter sun louver is used at south and southwest. West façade mainly uses stone curtain wall to effectively control the sunlight;

• Water-saving measures: All the toilets equip 6L water tanks; lavabos in public toilets use inductive faucets; urinals use inductive flush valves. Feed pump for life uses high-efficient energy-saving water pump to green the roof and collect the rainwater.

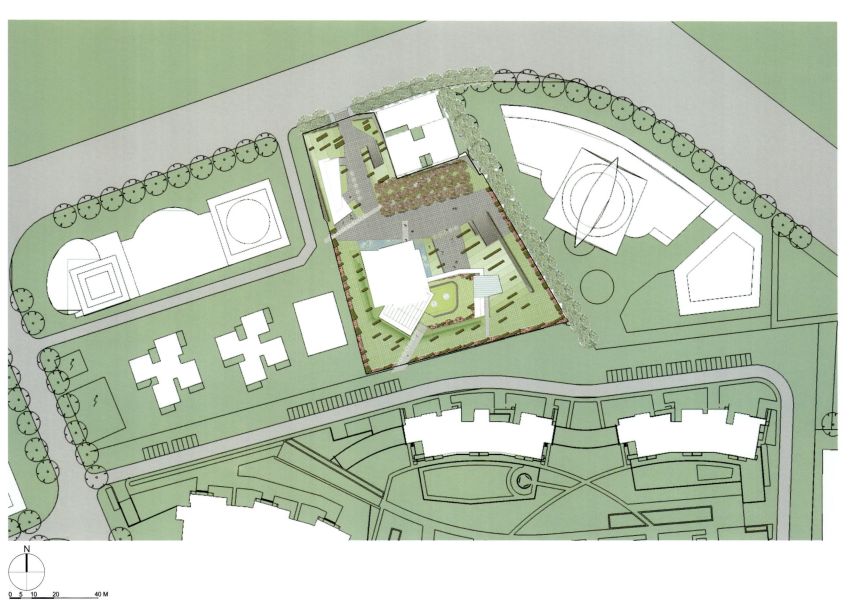

N

0 5 10   20   40 M

ARCHITECTURAL DESIGN FOR 1080 EAST DAMING ROAD PLAZA, SHANGHAI

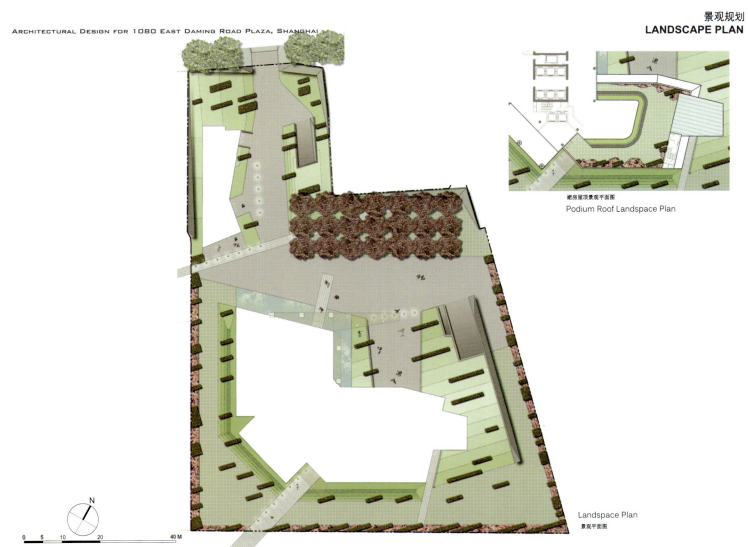

裙房屋顶景观平面图
Podium Roof Landspace Plan

Landspace Plan
景观平面图

N

0 5 10   20   40 M

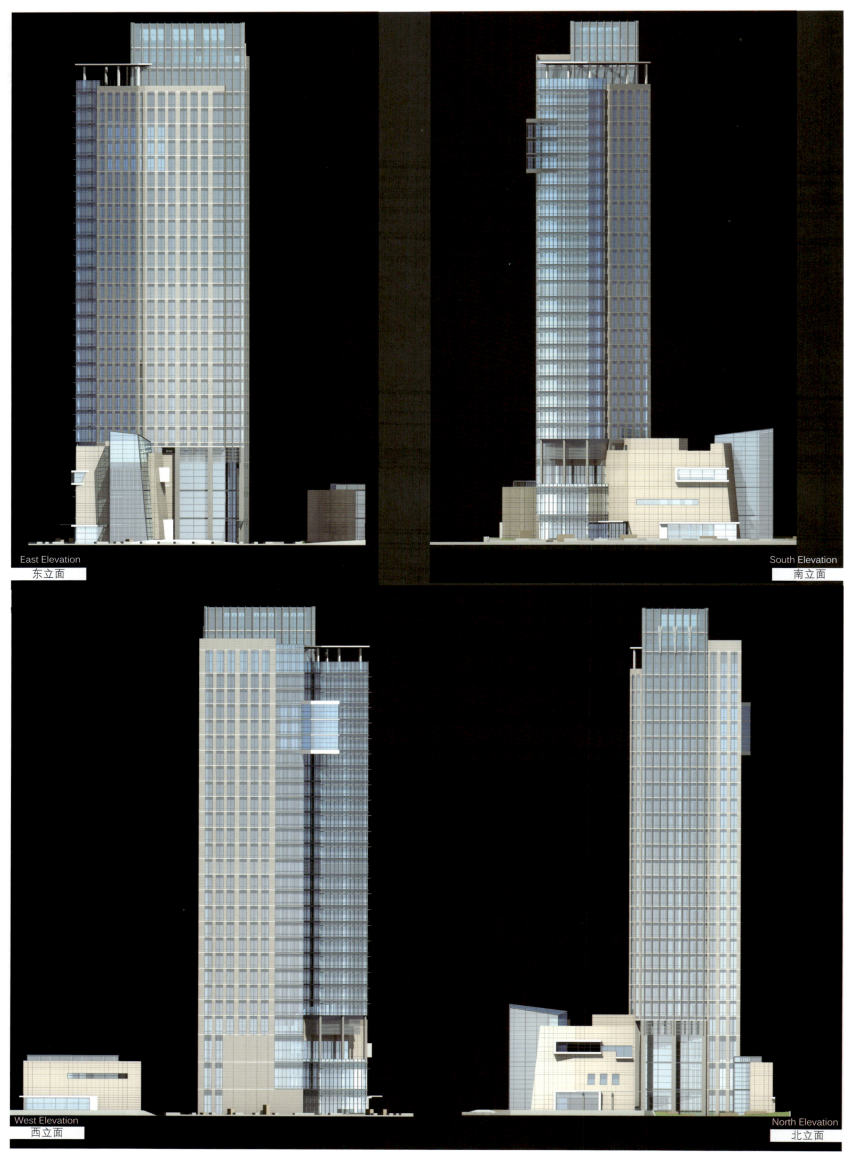

East Elevation
东立面

South Elevation
南立面

West Elevation
西立面

North Elevation
北立面

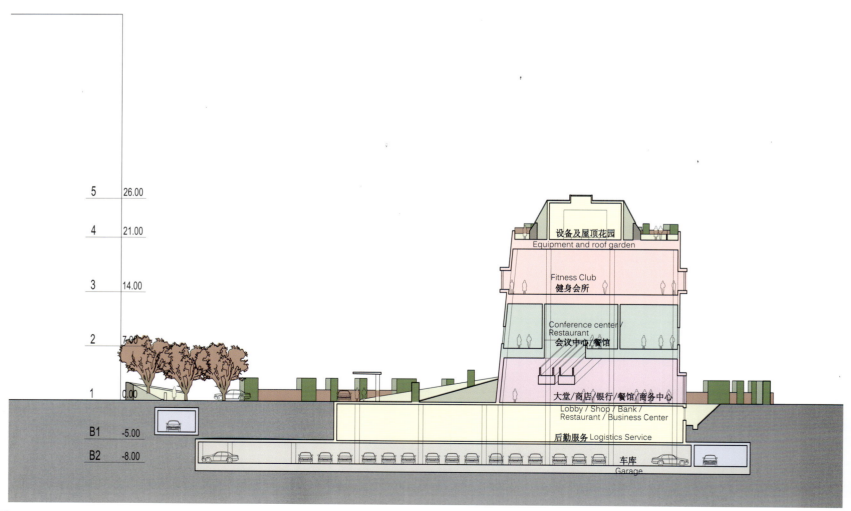

5	26.00			设备及屋顶花园 Equipment and roof garden
4	21.00			Fitness Club 健身会所
3	14.00			Conference center / Restaurant 会议中心 / 餐馆
2	7.00			
1	0.00			大堂 / 商店 / 银行 / 餐馆 / 商务中心 Lobby / Shop / Bank / Restaurant / Business Center
B1	-5.00			后勤服务 Logistics Service
B2	-8.00			车库 Garage

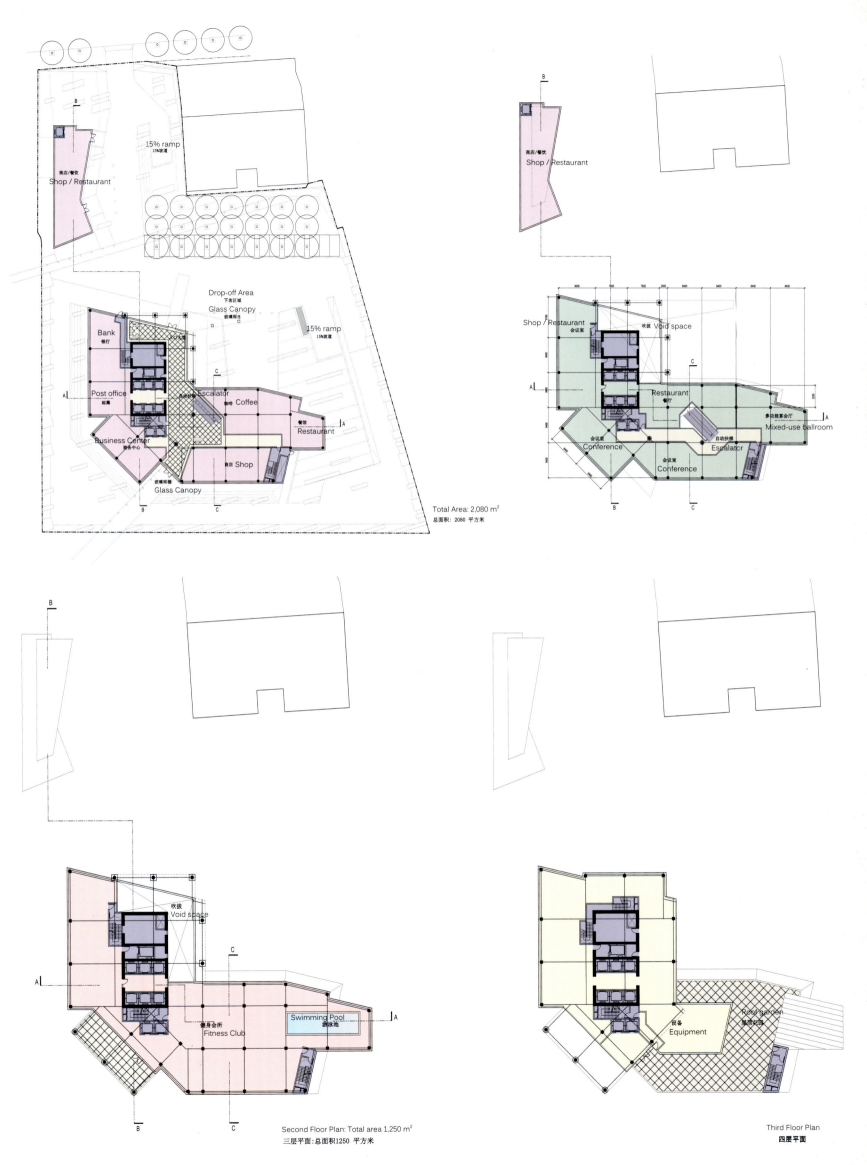

商店/餐饮
Shop / Restaurant

15% ramp
15%坡道

Drop-off Area
下客区域
Glass Canopy
玻璃雨店

15% ramp
15%坡道

Bank
银行

Post office
邮局

Escalator

Coffee
咖啡

Restaurant
餐馆

Business Center
商务中心

Glass Canopy
玻璃雨棚

Shop
商店

Total Area: 2,080 m²
总面积: 2080 平方米

商店/餐饮
Shop / Restaurant

Shop / Restaurant
会议室

吹拔 Void space

Restaurant
餐厅

多功能宴会厅
Mixed-use ballroom

会议室
Conference

自动扶梯
Escalator

会议室
Conference

吹拔
Void space

健身会所
Fitness Club

游泳池
Swimming Pool

设备
Equipment

屋顶花园
Roof garden

Second Floor Plan: Total area 1,250 m²
三层平面:总面积1250 平方米

Third Floor Plan
四层平面

205

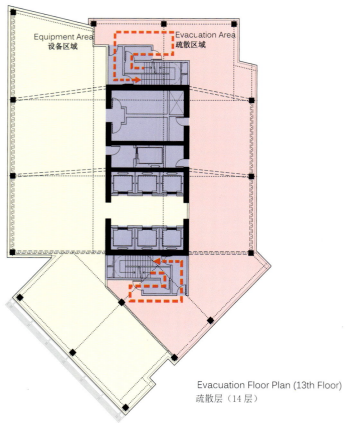

Evacuation Floor Plan (13th Floor)
疏散层（14层）

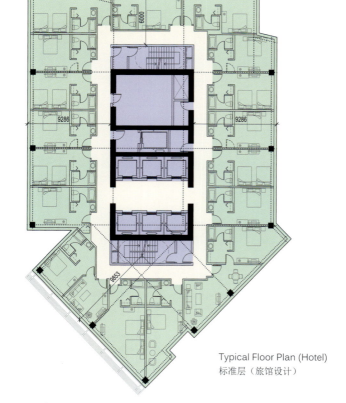

Typical Floor Plan (Hotel)
标准层（旅馆设计）

东方海港国际大厦是一栋多功能大厦，办公大楼高达 120 米，屋顶花园建在 21 米高的裙房上，并设有带有游泳池的健身俱乐部，内有会议厅和宴会厅，一楼为零售区，还有高达 10 米的独立购物中心。大厦的顶楼建有餐厅和阳台，在那里可以俯瞰城市天际线。项目采用室内环境可持续发展设计，例如，能源和水的保护措施、回收利用、可再生材料等，因此使其成为了中国上海第一栋获得能源与环境设计先锋奖金奖认证的高层建筑。

大厦的业主和设计团队的目标是建造一个多功能的、先进的地标性建筑，强调节

能环保，同时还要达到中华人民共和国的防火安全标准。

·暖通设计：复合式土壤源热泵系统，高效螺杆式冷水机组，热回收新风机组、VA 变风量系统（静压控制），水泵变频（冷却水、冷冻水、地源测水泵），冷却塔双速风机；

·CO_2 传感器控制系统 （按需通风）；

·过度季节自然通风，免费取冷；

·在建筑外窗区域设日光补偿的开关型传感器；

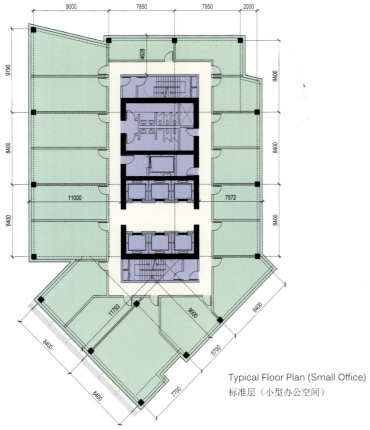

Typical Floor Plan (Small Office)
标准层（小型办公空间）

Typical Floor Plan (Mixed-use Office)
标准层（混合型办公空间）

· 优化照明密度 LPD，采用高效节能灯；

· 照明充分利用自然光，走道采用延时声光控开关。各房间照明按天然采光状况分区、分组控制；

· 人员感应器；

· 高效低能耗变压器，就地无功补偿措施，提供功率因数；

· 选用高性能玻璃，并且在各层设有可开启窗户，满足自然通风和换气，减少空调使用时间；

· 建筑外围护结构，玻璃幕墙使用 LOW-E 中空钢化玻璃。石材幕墙，50 毫米厚伊通轻质混凝土板，外保温一体化单元式幕墙体系；

· 采用合适的遮阳形式，有效控制眩光的产生。南向和西南向采用带有水平百叶遮阳板的玻璃幕墙。西立面以石材幕墙为主，达到对日照的有效控制；

· 节水措施：座便器均采用 6 升冲洗水箱，公共卫生间洗手盆均采用感应式龙头，卫生间小便器均采用感应式冲洗阀。生活给水泵采用高效节能水泵，用于绿化屋面及雨水收集。

TAINAN YUWEN LIBRARY

台南市裕文图书馆

EEWH CERTIFIED 台湾绿色建筑认证

SUSTAINABLE & GREEN FEATURES 绿色特征

- *Highly efficient sun shading system* ／ 充分的自然光利用，室内灯光自动化控制
- *Air - VRV (Variable Refrigerant Volume) air-condition system to reduce energy consumption* ／ 低流卫生器具，暖通空调系统提供热水

ARCHITECT
Malone Chang, Yu-lin Chen

FIRM
MAYU architects+

CLIENT
Tainan Municipal Library

PROJECT TEAM
Malone Chang & Yu-lin Chen (Architects), Kwantak AUYEUNG, Jin-de HSU (Project team), Dong-long WU (Construction supervision)

STRUCTURAL ENGINEER
Tien-Hun Engineering Consultant Inc.

INTERIOR DESIGN
MAYU architects+

LANDSCAPE DESIGN
MAYU architects+

LOCATION
Tainan, Taiwan, China

AREA
2,965 m² (Site), 1,350 m² (Building), 3,144 m² (Total Floor)

PHOTOGRAPHER
Guei-Shiang Ke, Yu-lin Chen

Yuwen Library is a critical part of ongoing municipal projects to expand the reading environment for the communities throughout Tainan City. In its particular location, the key design goals of Yuwen Library are to pull together surrounding public facilities by its unique façades, and expose its inner activities to the city by locating large concrete "windows" at building corners. We see this library as a generator of civic programs, and the children's library, facing the elementary school across the street, plays an important role.

Sustainability

Sun – On top of the lower concrete base we half-lodge a wood volume which contains collective human knowledge and clad in vertical wood louvers. The harsh direct sunlight is filtered and diffused by these louvers to provide comfortable interior ambiance. The combination of louvers and expansive glazing generates a transparent and universal space.

Air – VRV (Variable Refrigerant Volume) air-condition system is used in this project to achieve higher efficiency and increased controllability. Combined with Total Heat Recovery technology, optimal sustainability can be ensured.

Biological Diversity – Although located in the dense urban setting, we strive to protect the local biological system. During construction, every measure is taken to protect the surrounding indigenous Taiwanese flame gold trees. The third floor roof garden, on the level of treetops, becomes biological bridges for local species to travel across the site.

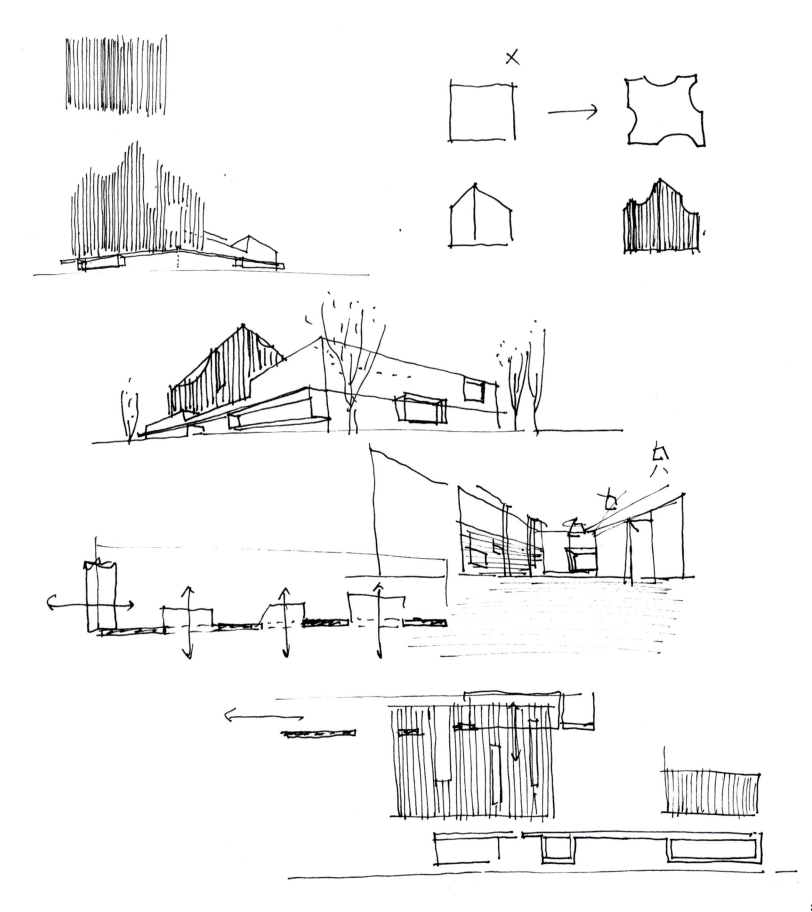

裕信路　Yu-xin Road

中山高速公路　freeway #1

裕文路　Yu-wen Road

N

0 20　　100　　　200
unit m

裕文图书馆是正在进行的市政项目中关键的一环，目的是为台南市居民扩大阅读环境。裕文图书馆地理位置独特，其最主要的设计目标就是通过独特的外墙立面将周围公用设施整合起来，拐角处的大型混凝土"窗"将建筑内部的情况展现给城市。我们将图书馆视为市民活动的发起者，少儿图书馆与小学隔街相望，意义重大。

可持续性

太阳——我们将半嵌入式的材积设计在凹陷的混凝土底座之上，这一设计包含了人类集体的智慧。同时木质百叶窗将刺眼的直射光过滤分散，营造出舒适的室内环境。百叶窗和膨胀玻璃相结合，共同创造出通透宽敞的空间。

空气——该项目使用变制冷剂流量空调系统，以获得更高的效率，增加可控性。加上全热回收技术的应用，最大限度地确保可持续性。

生物多样性——尽管坐落于密集的城区，我们仍致力于保护当地的生态系统。在建设期间，我们每一项措施都以保护中国台湾本土的火焰黄金树为前提。三楼的屋顶花园与树顶齐平，成为了当地物种和外来物种之间的生态桥梁。

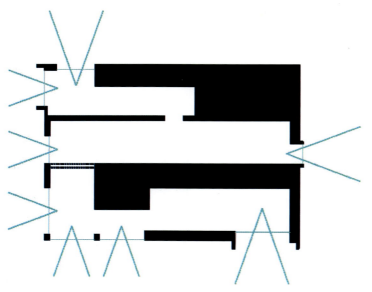

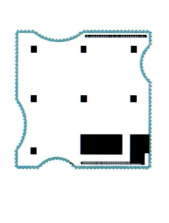

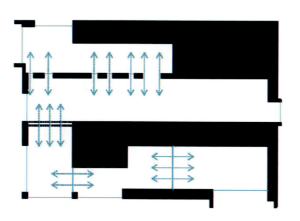

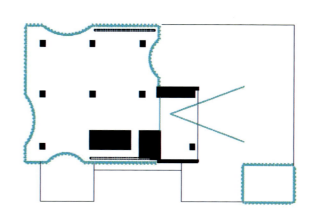

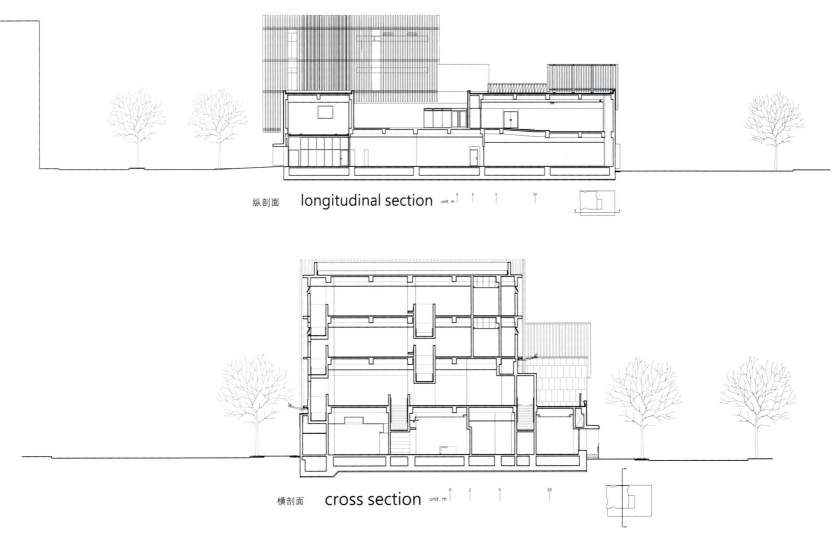

纵剖面　longitudinal section　unit：m

横剖面　cross section　unit：m

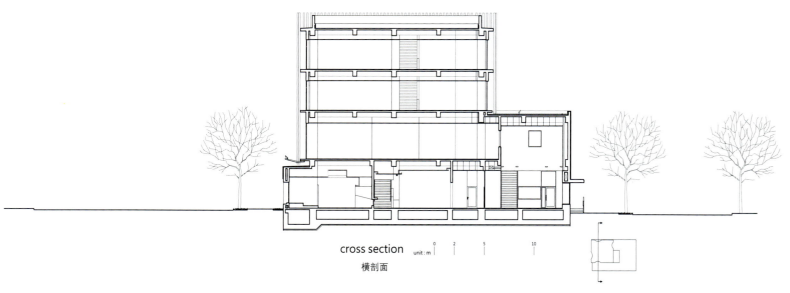

cross section   unit : m   0 2 5   10

横剖面

东立面　east elevation　unit : m　0　2　5　10

西立面　west elevation　unit : m　0　2　5　10

library

北立面　north elevation　unit : m　0　2　5　10

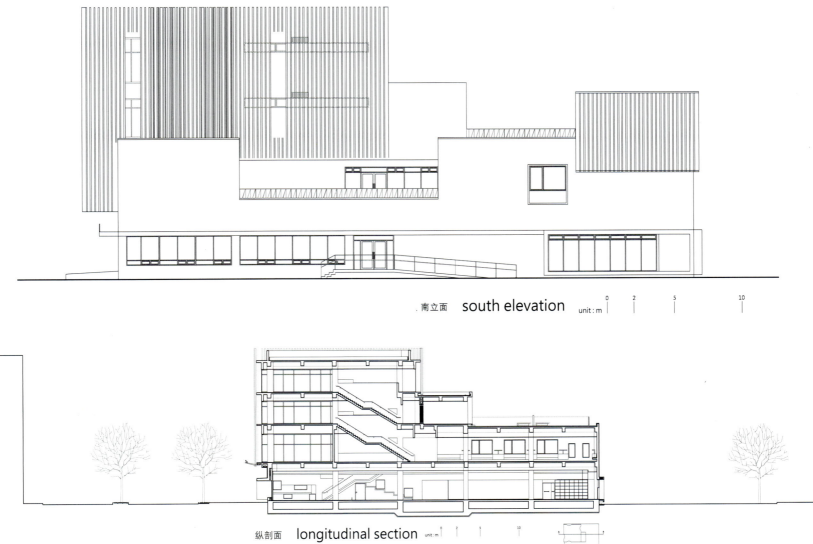

南立面　south elevation　unit : m

纵剖面　longitudinal section　unit : m

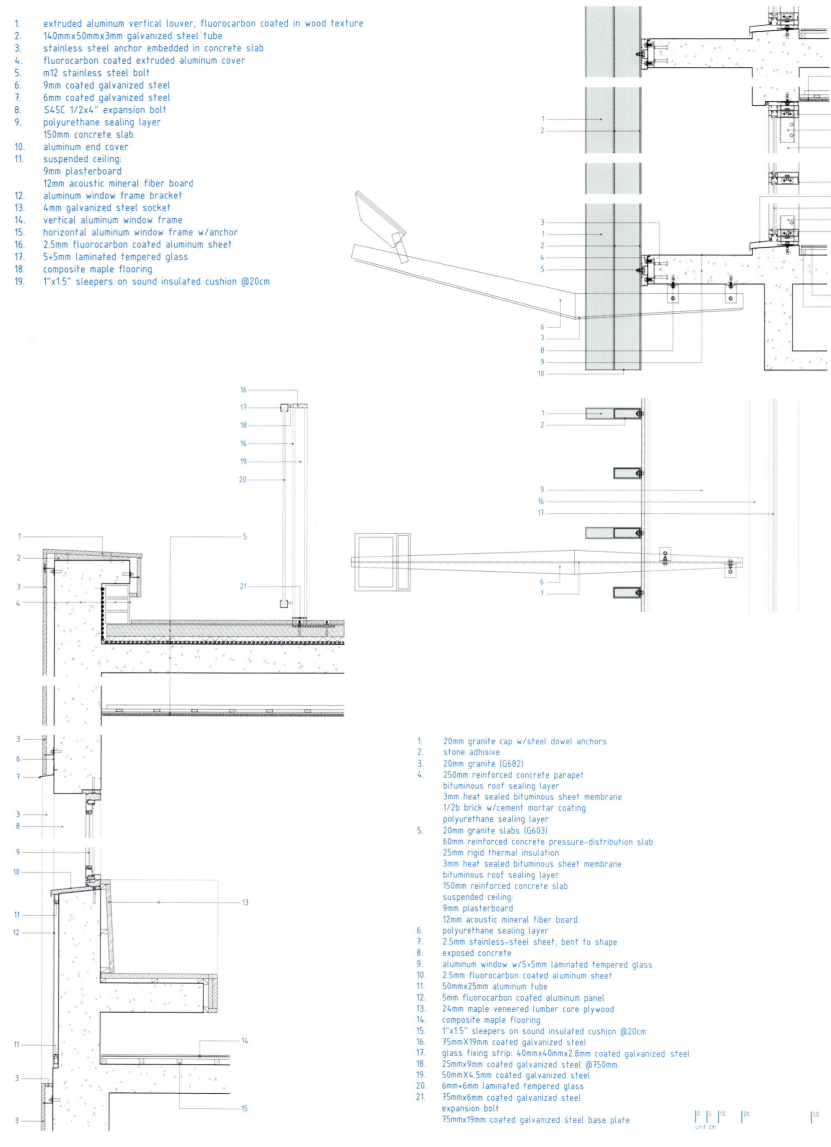

1. extruded aluminum vertical louver, fluorocarbon coated in wood texture
2. 140mmx50mmx3mm galvanized steel tube
3. stainless steel anchor embedded in concrete slab
4. fluorocarbon coated extruded aluminum cover
5. m12 stainless steel bolt
6. 9mm coated galvanized steel
7. 6mm coated galvanized steel
8. S45C 1/2x4" expansion bolt
9. polyurethane sealing layer
   150mm concrete slab
10. aluminum end cover
11. suspended ceiling:
    9mm plasterboard
    12mm acoustic mineral fiber board
12. aluminum window frame bracket
13. 4mm galvanized steel socket
14. vertical aluminum window frame
15. horizontal aluminum window frame w/anchor
16. 2.5mm fluorocarbon coated aluminum sheet
17. 5+5mm laminated tempered glass
18. composite maple flooring
19. 1"x1.5" sleepers on sound insulated cushion @20cm

1. 20mm granite cap w/steel dowel anchors
2. stone adhisive
3. 20mm granite (G682)
4. 250mm reinforced concrete parapet
   bituminous roof sealing layer
   3mm heat sealed bituminous sheet membrane
   1/2b brick w/cement mortar coating
   polyurethane sealing layer
5. 20mm granite slabs (G603)
   60mm reinforced concrete pressure-distribution slab
   25mm rigid thermal insulation
   3mm heat sealed bituminous sheet membrane
   bituminous roof sealing layer
   150mm reinforced concrete slab
   suspended ceiling:
   9mm plasterboard
   12mm acoustic mineral fiber board
6. polyurethane sealing layer
7. 2.5mm stainless-steel sheet, bent to shape
8. exposed concrete
9. aluminum window w/5+5mm laminated tempered glass
10. 2.5mm fluorocarbon coated aluminum sheet
11. 50mmx25mm aluminum tube
12. 5mm fluorocarbon coated aluminum panel
13. 24mm maple veneered lumber core plywood
14. composite maple flooring
15. 1"x1.5" sleepers on sound insulated cushion @20cm
16. 75mmX19mm coated galvanized steel
17. glass fixing strip: 40mmx40mmx2.8mm coated galvanized steel
18. 25mmx9mm coated galvanized steel @750mm
19. 50mmX4.5mm coated galvanized steel
20. 6mm+6mm laminated tempered glass
21. 75mmx6mm coated galvanized steel
    expansion bolt
    75mmx19mm coated galvanized steel base plate

unit: cm

216

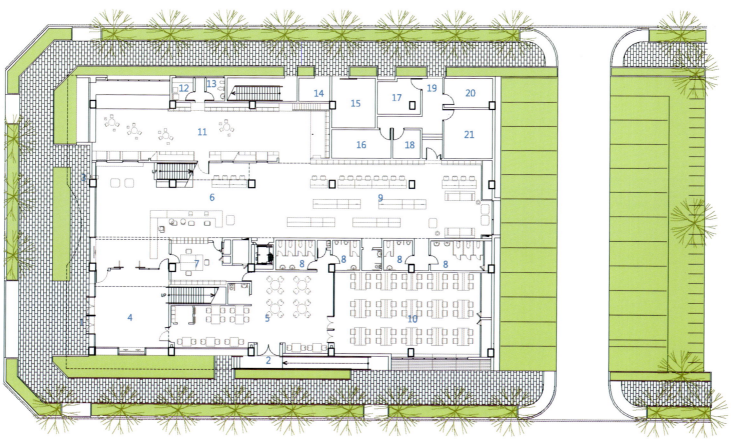

1	主入口	**main entrance**	8	厕所	**restroom**	15	台电配电室 **electricity distribution facilities**
2	次入口	**side entrance**	9	学龄儿童及青少年图书馆 **children&teenager library**	16	消防机械室 **firefighting pump room**
3	还书箱	**book return**	10	自修室	**reading area**	17	变电室 **transformer room**
4	入口大厅（上部挑空） **entrance lobby**	11	学龄儿童及青少年图书馆 **children&teenager library**	18	电信室 **telecommunication facilities**
5	社区书房（阅报区）**community reading room**	12	育婴室	**nursing room**	19	维修入口 **service entrance**
6	流通柜台及新书区 **circulation counter**	13	亲子厕所 **parent-child restroom**	20	发电机室 **generator room**
7	工作及志工休息区 **office**	14	垃圾分类储存空间 **recycling storage**	21	机房 **mechanical facilities**

1F plan

N 0 1 5 10
unit m

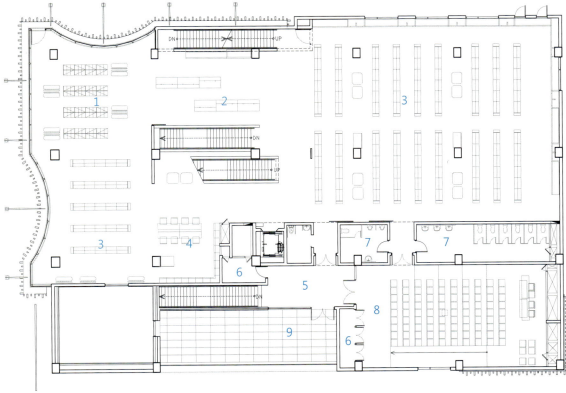

2F plan

1	现期期刊区 periodicals	6	储藏室 storage
2	主题展示区 special display	7	厕所 restrooms
3	书库及阅读区 book stacks/ reading area	8	多功能室 Multi-purpose room
4	参考书区 reference book	9	屋顶花园 roof garden
5	室内廊道 foyer		

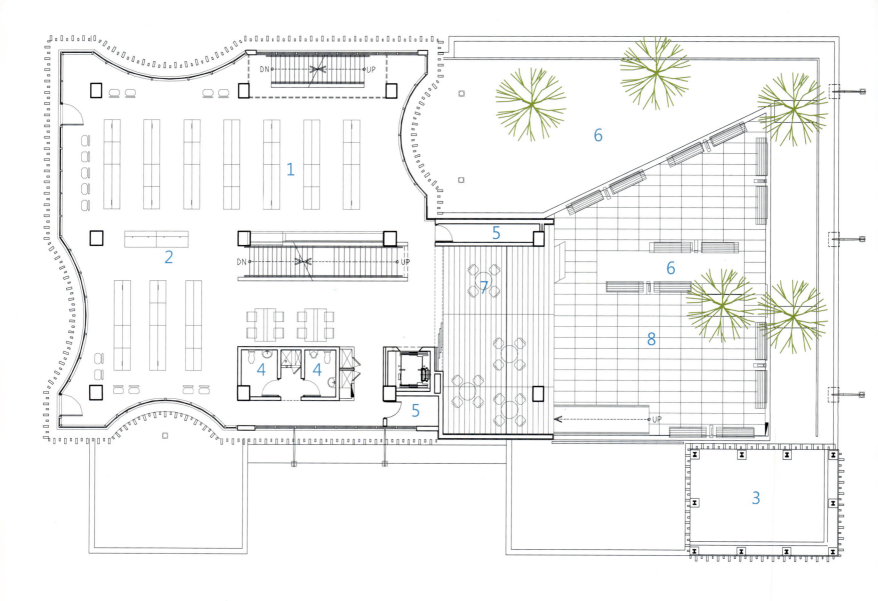

1	书库及阅读区 book stacks/ reading area	5	储藏室 storage
2	主题展示区 special display	6	公共艺术区域 public art area
3	空调主机 AC	7	半户外阅读空间 semi-outdoor reading area
4	厕所 restrooms	8	屋顶花园 roof garden

3F plan

N

0 1 5

unit m

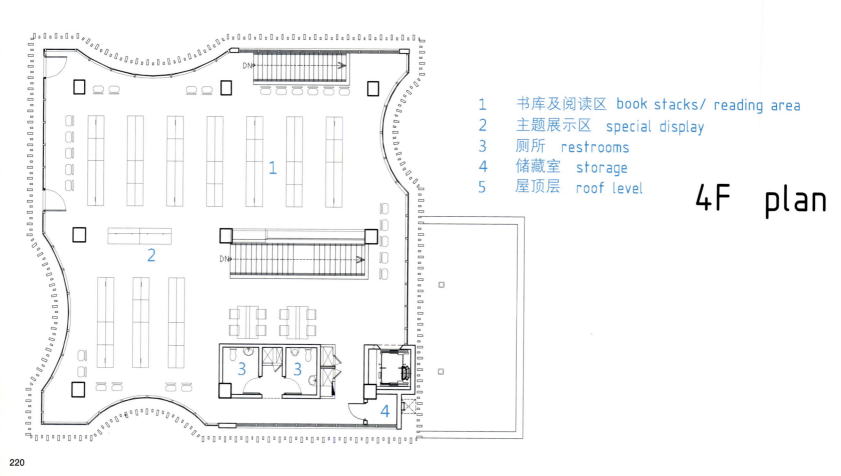

1	书库及阅读区 book stacks/ reading area
2	主题展示区 special display
3	厕所 restrooms
4	储藏室 storage
5	屋顶层 roof level

4F plan

TAIPEI FLORA EXPO PAVILIONS 花博新生公园三馆

WORLD GOLD AWARD FOR DESIGN AND CONSTRUCTION OF THREE-DIMENSIONAL GREENING

世界立体绿化设计施工金奖

ARCHITECT
Bio-architecture formosana

LOCATION
Taipei, Taiwan, China

AREA
161,057 m² (Site Area), 14,391 m² (Gross Floor Area), 8,643 m² (Built Area)

SUSTAINABLE & GREEN FEATURES 绿色特征

- *Use of local plants to reduce energy consumption* ／ 使用当地植物降低能源消耗
- *Large use of solar panels* ／ 大量使用太阳能板
- *Green roof to recycle rainwater and lighten heat island effect* ／ 草坪屋顶用于雨水回收，减轻热岛效应

Taipei Flora Expo Pavilions (dream, future, life) represents our long-time effort of integrating architectural art and technology. This project turns the old trees in New Park base into a magic driving force inviting buildings to extend, either high or low, among them, offering visitors scenic view and communication. Such spatial experience is like no other in Taiwan. The building meets the diamond standard of the interior green architecture of Taiwan Ministry of Interior. Furthermore, Intelligence technology is found in the control of all the windows, sunshade, temperature and air conditioning system. Thus the comfort and energy consumption of the inner building could be in optimum state at any moment.

An aerial view of Taipei Flora Expo Pavilions base shows groups of trees and two pieces of large space. During the planning, the sponsor has realized in advanced the necessity of reducing the number of transplanting trees in park. During the construction, we strictly required the company concerned that they should not take those old trees as "parking shad" in order to protect the soil and the roots from the constant rolling of tires. We also realized that buildings might help trees disperse the pressure of northeast wind and typhoon, reduce the hours of direct sunlight and increase the survival rate.

When designing the three pavilions, many domestic and foreign technologies were fused. For example, Future Pavilion itself is a huge greenhouse which houses Taiwan native plants from frigid, temperate and tropical zones. This was made possible by high-tech temperature and humidity control system. For the plants from subtropics zone and orchid zone, water walls and air fans are used to take the place of energy-consuming air conditioning system. The roof is also the result of high-tech. Greenhouse uses ETFT cushion which is transparent but insulated. It helps with the plants' photosynthesis. Curtain design can adjust the indoor light according to the weather. In addition, the 200KW solar panel could provide alternative energy and reduce the carbon emission.

The space requirement mode can be controlled automatically, and energy is saved and temperature is lowered by the drencher curtain in the greenhouse. The hall is equipped with mist spray and floor cooling instead of air-conditioner. Light green energy-saving roof softens Taipei's skyline and reduces urban heat island effect. Plants on the wall can mean low maintenance and thermal insulation. Solar panel on the inclining façade toward the courtyard provides recycling energy.

We are trying to figure out whether we could reduce the negative effects meanwhile positively make our lives more elegant and healthy. After all, the Expo is temporary while the environment is eternal. Hopefully after the Flora Expo, our five senses could experience fun in life brought by the green and water, enjoying moments with our family in the park.

绿建筑计划说明图
Green building design illustration

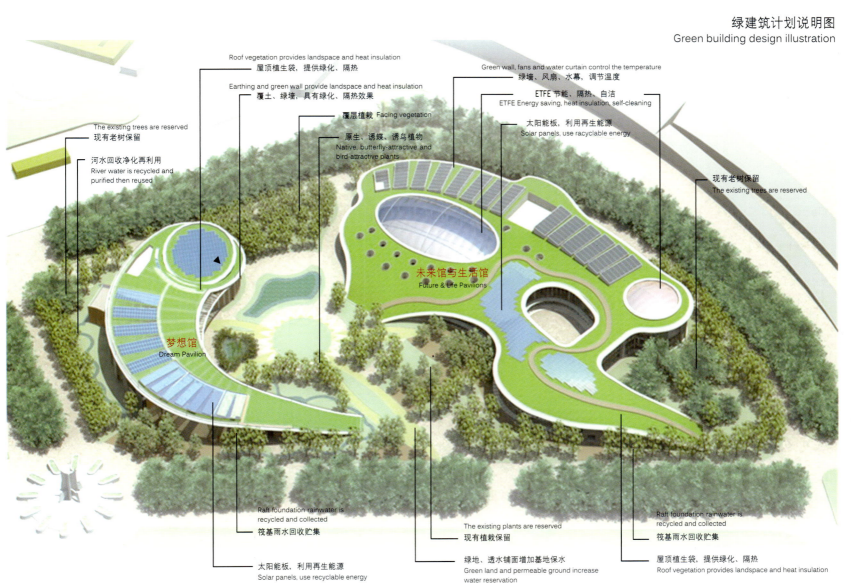

Roof vegetation provides landspace and heat insulation
屋顶植生袋，提供绿化、隔热

Green wall, fans and water curtain control the temperature
绿墙、风扇、水幕，调节温度

Earthing and green wall provide landspace and heat insulation
覆土、绿墙，具有绿化、隔热效果

ETFE 节能、隔热、自洁
ETFE Energy saving, heat insulation, self-cleaning

覆层植栽 Facing vegetation

太阳能板，利用再生能源
Solar panels, use racyclable energy

原生、诱蝶、诱鸟植物
Native, butterfly-attractive and bird-attractive plants

The existing trees are reserved
现有老树保留

河水回收净化再利用
River water is recycled and purified then reused

现有老树保留
The existing trees are reserved

未来馆与生活馆
Future & Life Pavilions

梦想馆
Dream Pavilion

Raft foundation rainwater is recycled and collected
筏基雨水回收贮集

The existing plants are reserved
现有植栽保留

Raft foundation rainwater is recycled and collected
筏基雨水回收贮集

太阳能板，利用再生能源
Solar panels, use recyclable energy

绿地、透水铺面增加基地保水
Green land and permeable ground increase water reservation

屋顶植生袋，提供绿化、隔热
Roof vegetation provides landspace and heat insulation

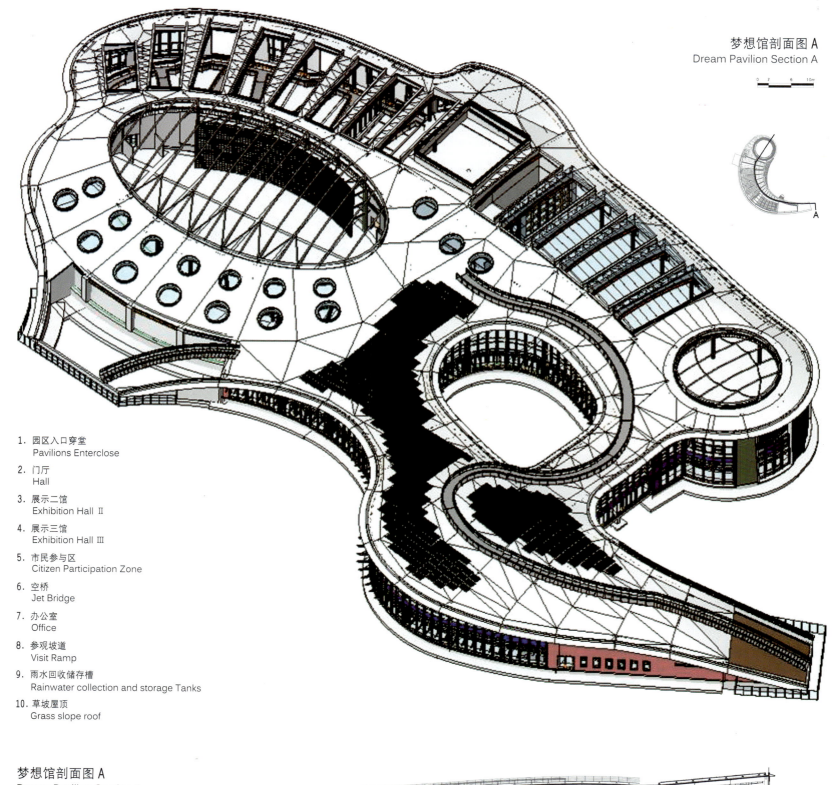

A

1. 园区入口穿堂
   Pavilions Enterclose

2. 门厅
   Hall

3. 展示二馆
   Exhibition Hall Ⅱ

4. 展示三馆
   Exhibition Hall Ⅲ

5. 市民参与区
   Citizen Participation Zone

6. 空桥
   Jet Bridge

7. 办公室
   Office

8. 参观坡道
   Visit Ramp

9. 雨水回收储存槽
   Rainwater collection and storage Tanks

10. 草坡屋顶
    Grass slope roof

梦想馆剖面图 A
Dream Pavilion Section A

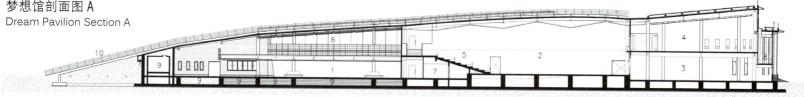

　　花博新生公园三馆（梦想馆、未来馆、生活馆）整体呈现了我们长久以来在整合建筑艺术和科技方面的努力。此案将新生公园基地内的老树，化为驱动空间和动线的魔术师，让建筑在树群之间伸展，或低伏或扬起，引导参观者的视线，使其与空间交流，是台湾地区少有的空间经验。本建筑除了达到台湾内政部绿色建筑标章钻石级的标准外，还利用智能技术控制所有的窗户、遮阳、室温控设备、空调系统，因此将建筑内部的舒适与节能随时调整在最佳状态。

　　若从空照图看花博新生公园三馆的基地，你将看到一片片的树木和两片较大的空地。在规划中，主办单位已预先看到降低移植公园现有树木数量的必要性。在施工期间，我们严格要求施工单位，不得把保留的老树或树群当成"停车棚"，以避免轮胎不断滚压，破坏土壤的活性和树木根系的活力。我们还体悟到建筑可以帮树木分散强劲的东北风、台风的压力，并能减少直接日照的时数，提高树木缺水期的存活率。

　　这三座展馆的设计融合了很多国内外的技术。例如，未来馆本身就是一栋巨大的温室，在高科技的温度和湿度控制系统的协助下，种植了寒带、温带、热带的台湾原生植物；而亚热带区及兰花区，则用水墙与风扇代替较耗能的空调，以维持最适合植物生长的环境。屋顶也是高科技的结晶。温室采用 ETFE 气枕，其隔热却透光的特性，让室内植物能进行光合作用；遮帘设计能随天气变化，调整室内光亮度；此外，200千瓦的太阳能板，则提供了替代能源，可减少碳排放量。

　　该建筑的空间需求模式可由机械自动控制，而温室的水帘则有节能、调温的作用。大厅设置了喷雾并引进了地冷设备，为无空调设计。轻质绿化的节能屋顶，软化了台北的天际线，降低了都市的热岛效应。墙壁绿化围护成本较低，并可隔热，向中庭倾斜的屋面装置太阳能则为建筑提供了再生能源。

　　我们正在思考让建设在减少负面的影响的同时，尽可能地让人们生活得更优雅、健康。毕竟，博览会仅是一时的，但环境却是永远的。或许在花博会结束后，我们的五感能透过公园的转变而体验到一种满眼绿色、水流环境的生活趣味，享受一家人到公园休闲的乐趣。

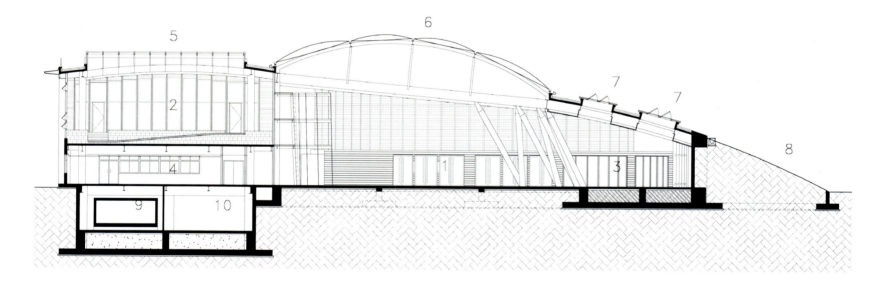

1. 未来馆（热带／亚热带植物展示温室）
Future Pavilion (Tropical/Subtropical Plant Show Greenhouse)

2. 未来馆（温带植物展示温室）
Future Pavilion (Temperate Plant Show Greenhouse)

3. 未来馆（耐荫植物展示温室）
Future Pavilion (Shade-tolerant Plant Show Greenhouse)

4. 办公室
Office

5. 温室天窗＋内外遮荫网
Greenhouse Skylight + Inside and Outside Shade Net

6. ETFE 膜构造
ETFE Membrane Structure

7. 温室天窗
Greenhouse Skylight

8. 草坡屋顶
Grass Slope Roof

9. 水箱
Water Tank

10. 机房
Machine Room

未来馆与生活馆剖面图 A
Future & Life Pavilions Section A

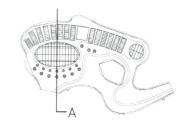

NANTONG URBAN PLANNING MUSEUM

南通市通州区城市展览馆

SUSTAINABLE & GREEN FEATURES 绿色特征

• *Double-layer façade: interior glass maintenance structure to reduce heat loss, parametric-design sun-shading system to make a breathing exterior façade* ／ 双层表皮系统：内层的玻璃维护结构减少热量流失，参数化设计的遮阳系统形成一面可呼吸的外墙

ARCHITECT AND LEAD CONSULTANTS
HENN Architekten

PROJECT PRINCIPAL
Gunter Henn

PROJECT TEAM
Leander Adrian, Daniel da Rocha, Martin Henn, Anthony
Hu, Alan Kim, Agata Kycia, Paul Langley, Jeewon Paek,
Emil Pira, Klaus Ransmayr, Wei Sun, Mu Xingyu

PLANNING
Georg Pichler, Arjan Pit, Mu Xingyu, Jakob Drömmer

LANDSCAPE DESIGN
HENN Architekten

CLIENT
Urban Planning Department of Nantong, Tongzhou

LOCATION
Nantong, Jiangsu Province, China

AREA
8,100 m² (Gross Floor), 7,000 m² (Net
Floor), 4,900 m² (Exhibition)

PHOTOGRAPHER
Bartosz Kolonko

As part of the master plan designed by HENN Architekten, the new Nantong Urban Planning Museum is located prominently along the central river, whose course, together with existing cultural and commercial facilities, establish the primary East-West axis of Nantong.

The museum is characterized as a 16m, floating volume, which rests on a setback glass pedestal, offering space for special exhibitions, café and bookstore. The overall dominant form which cantilevers above the glass entry, contains the primary exhibition space, offices and conference rooms.

Its distinctive façade is composed of two layers: the inner which thermally seals the building envelope, and the outer, a reticulated metal structure with a gradient of varied panels. The façade's diamond-shaped Diagrid is comprised of seven different panels that allow for varying degrees of opening from 9% to 60%. This provides an opportunity for the controlled regulation of sunlight in fine increments, to accommodate the needs of the interior program.

The exhibition spaces are therefore, characterized by a predominantly closed façade with minimal openings, and the offices with maximum levels of natural daylight.

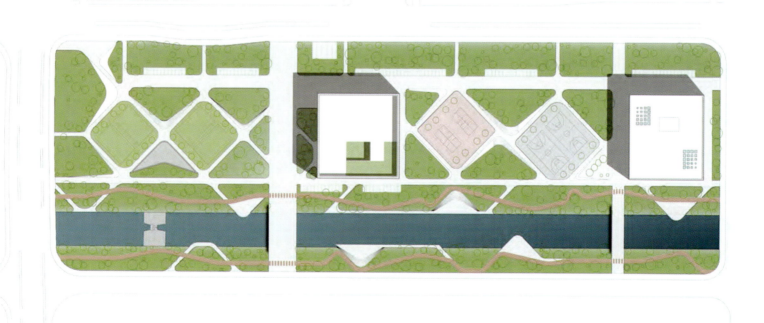

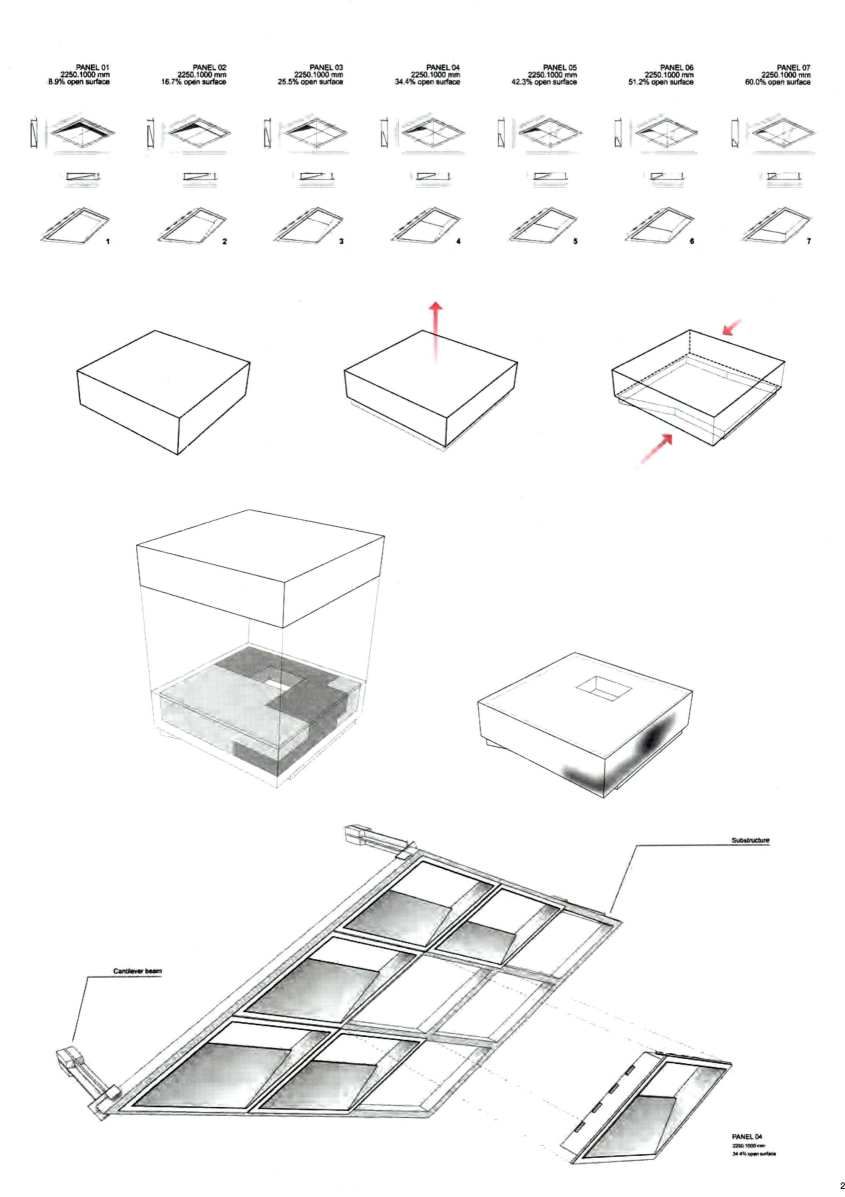

PANEL 01
2250.1000 mm
8.9% open surface

PANEL 02
2250.1000 mm
16.7% open surface

PANEL 03
2250.1000 mm
25.5% open surface

PANEL 04
2250.1000 mm
34.4% open surface

PANEL 05
2250.1000 mm
42.3% open surface

PANEL 06
2250.1000 mm
51.2% open surface

PANEL 07
2250.1000 mm
60.0% open surface

1

2

3

4

5

6

7

Substructure

Cantilever beam

PANEL 04
2250.1000 mm
34.4% open surface

SOUTH 南

EAST 东

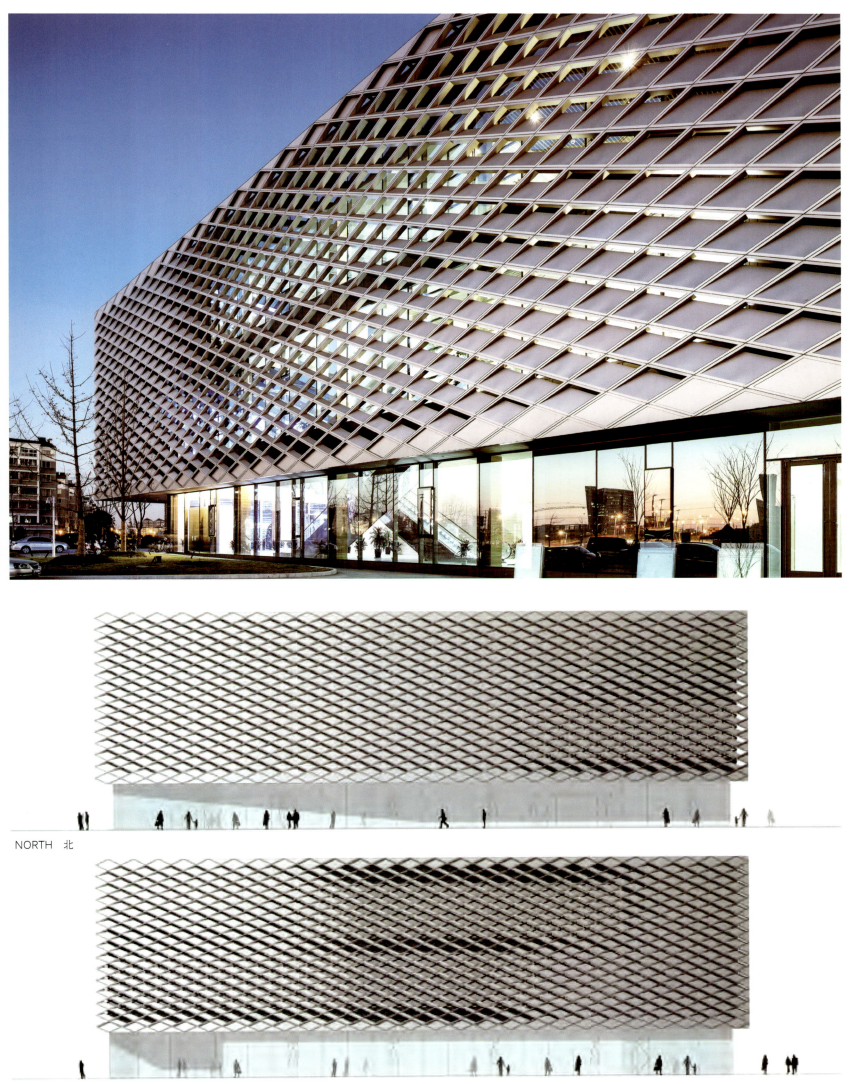

NORTH 北

WEST 西

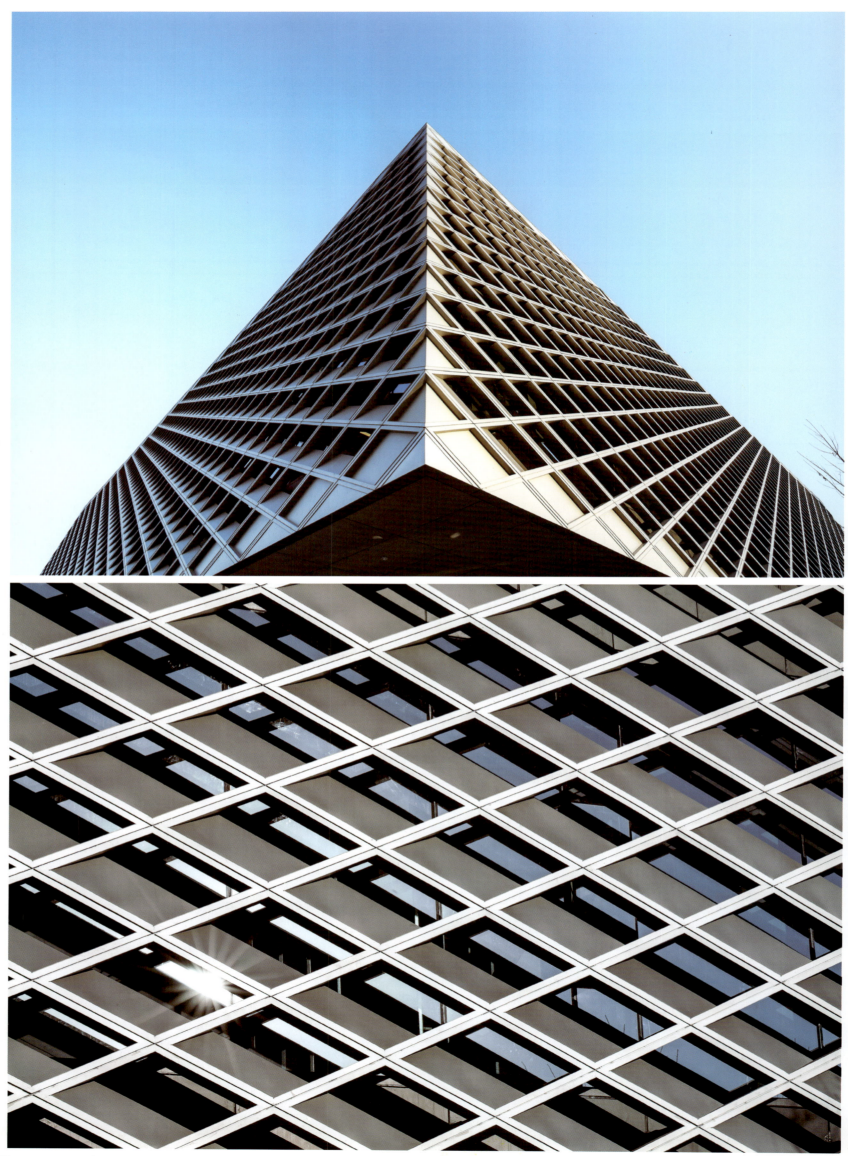

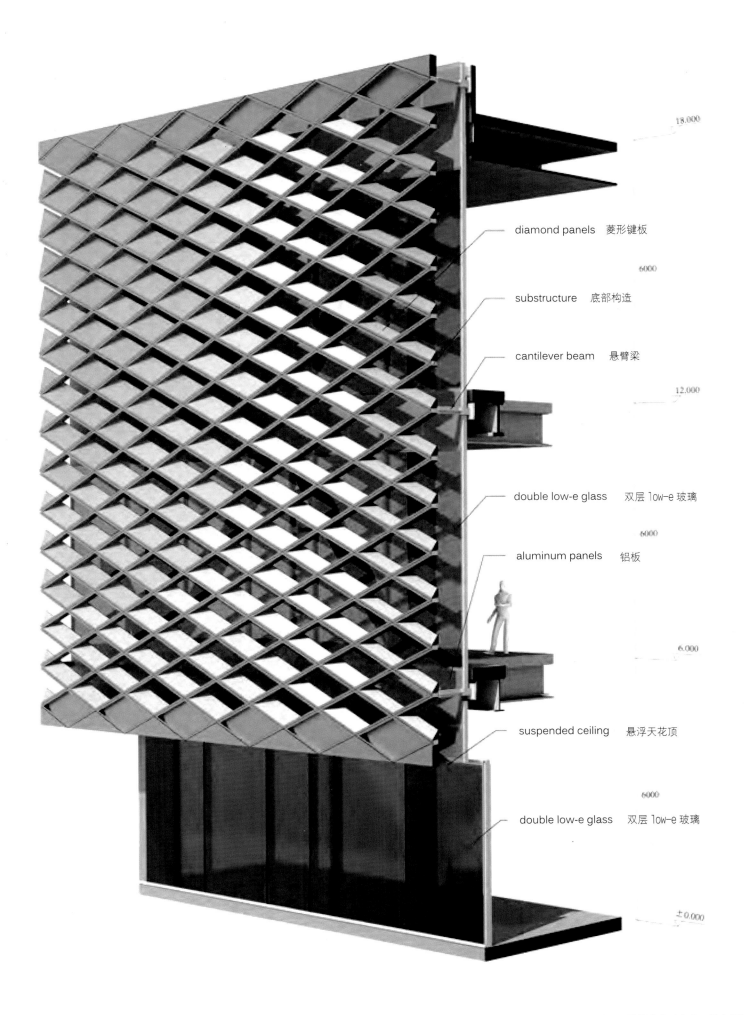

diamond panels 菱形键板

substructure 底部构造

cantilever beam 悬臂梁

double low-e glass 双层 low-e 玻璃

aluminum panels 铝板

suspended ceiling 悬浮天花顶

double low-e glass 双层 low-e 玻璃

18.000

6000

12.000

6000

6.000

6000

±0.000

新建的南通市通州区城市展览馆是 HENN Architekten 设计的总体规划的一部分，毗邻中心河流，该河的水道与现存的文化及商业设施共同确立了南通市主要的东西向中轴线。

该博物馆是一个高达 16 米的悬空体量，支撑在旋涡状的玻璃底座上，那里有专题展览厅、咖啡厅及书店。博物馆利用悬臂支撑于玻璃入口之上，里面包括主要的展览空间、办公区域及会议室。

其外观独具特色，由两层组成：内层将建筑围护结构牢牢包住，具有保温的作用；外层是金属网状结构，各式嵌板形成了一个倾斜的角度。建筑立面的斜肋构架呈菱形，由七种不同的嵌板组成，张开度为 9% ~ 60% 不等。这样一来，就可以准确地调节阳光进入量，从而适应内部使用的需要。

因此，展区有这样一个特点，即立面开口极小，基本上是封闭的，而办公室却可以最大程度地享受自然光。

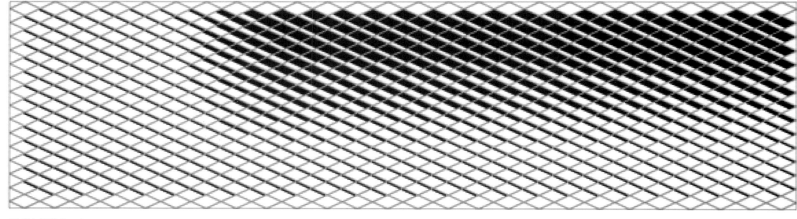

SOUTH 南

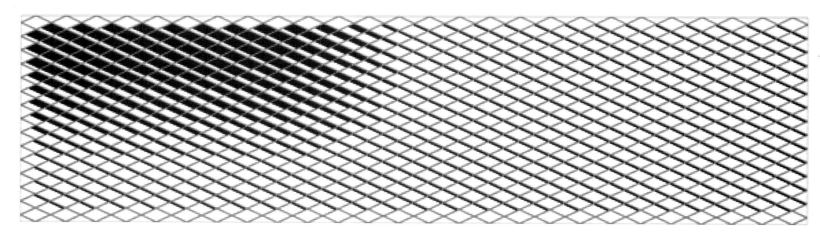

EAST 东

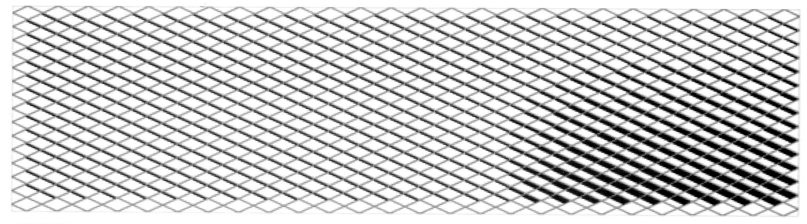

NORTH 北

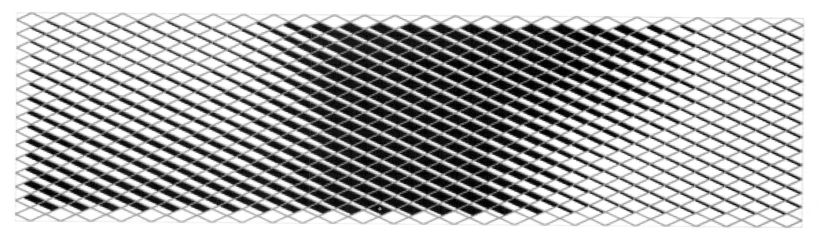

WEST 西

242

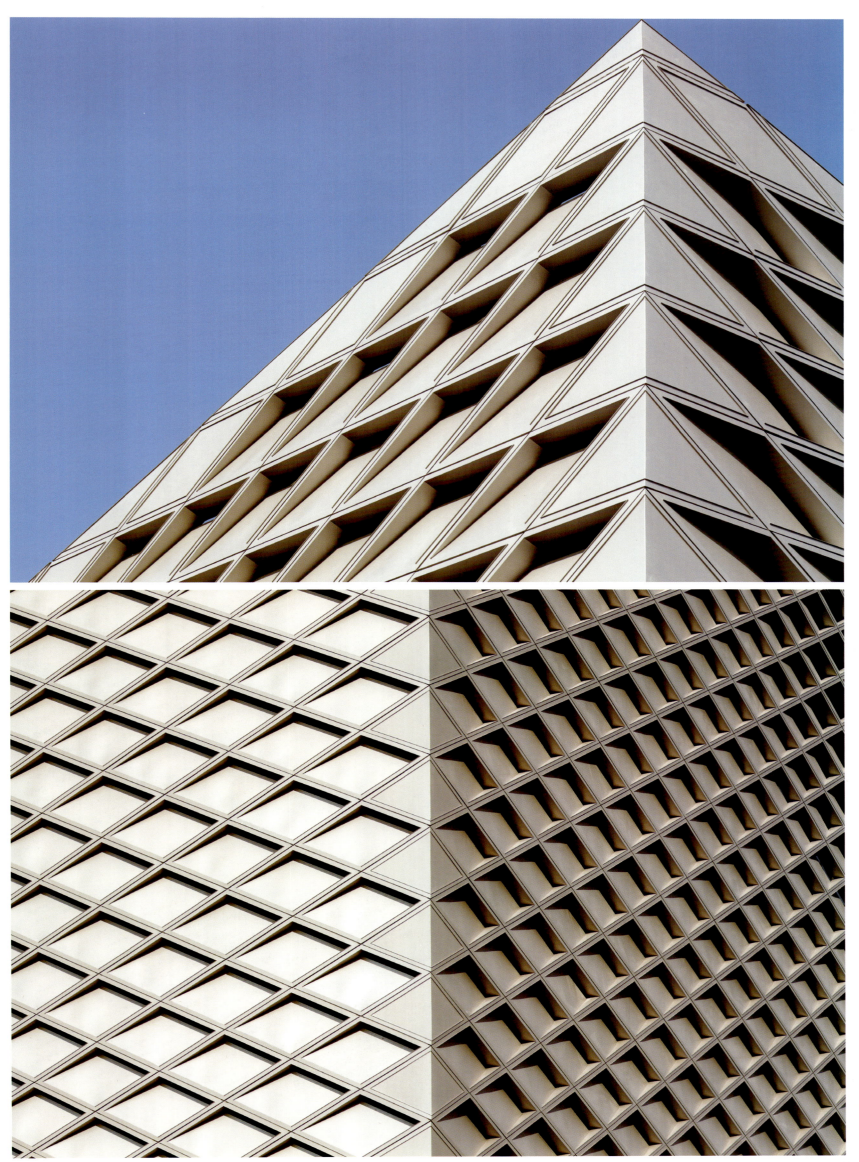

SUZHOU INDUSTRIAL PARK
MERCHANT BANK BUILDING

苏州工业园区
中国招商银行大厦

• *Highly efficient double-layer Low-E glass curtain wall to maximize natural ventilation and reduce energy consumption* ／ 高效的双层 Low-E 玻璃幕墙系统最大限度地使用自然通风，降低能源消耗

• *Roof garden to recycle rainwater and lighten the heat island effect* ／ 屋顶用于回收雨水，减轻热岛效应

ARCHITECT
Jeff Walker, Carol Zhang,

FIRM
Johnson Pilton Walker

PROJECT LEADER
Dickson Leung

PROJECT TEAM
Supinder Matharu, Nigel Carusi, Adam Rusan

MECHANICAL, ELECTRICAL, HYDRAULIC AND FIRE ENGINEERING AND ARCHITECTURAL CONSTRUCTION DOCUMENTATION
Suzhou Institute of Architectural Design Co. Ltd.

INTERIOR
Shanghai Hanshi Interior Design Co Ltd

CURTAIN WALL DESIGN
Shanghai Tongyi Design Co Ltd

BUILDER
Nantong No. 4 Construction Company

PILING AND FOUNDATION CONTRACTOR
Suzhou Zhong Jian Foundation Company

CONSTRUCTION SUPERVISION
SIP Construction Supervision Pty Ltd

LIGHTING DESIGN AND CONSTRUCTION
Yue Xiu Lighting Pty Ltd

CURTAIN WALL CONSTRUCTION
Kelida Building and Decoration Co Ltd

LOCATION
Suzhou, Jiangsu Province, China

AREA
30,000 m²

PHOTOGRAPHER
Yao Li

The building presents a distinctive and identifiable form within the precinct. Two slender planes of polished granite appear to support a glazed tower over a podium and frame the principal entry. The regular grid of recessed 1m x 1m windows, many of which open for natural ventilation, creates a unique ambiguous sense of scale, and belies the building's height which is limited to 100m by the planning controls.

At close quarters the granite planes appear as crisply detailed screens. Traditional Chinese architecture uses screens to frame views, filter sunlight and importantly, provide a sense of privacy whilst maintaining a strong connection to the outside. The granite screens of the China Merchant Bank achieve the same effect at both a precinct and human scale, with a sense of permanence and privacy that is so critical for a bank building.

The western façade of the suspended glass box is a clear low-E glazed dual wall system, with interconnecting stairs between floors set within a wider section of the winter garden which also functions as an informal meeting space. These circulation elements, uncommon in many Chinese commercial buildings, have proved extremely successful in fostering staff interaction and a sense of community and in changing the traditional rigid workplace environment.

The dual wall system enables extensive areas of the building to be naturally ventilated and significantly reduces the building's energy consumption.

With extensive areas of natural ventilation, much of which affords great individual user control, a highly efficient dual wall façade and a podium roof that captures and reuses rainwater, the China Merchant Bank building is an environmentally sensitive response to a challenging climate, functional brief and budget.

Importantly, the building also makes a significant contribution to social sustainability by providing a range of planning and functional innovations, such as interconnecting stairs and break out spaces, which have been very successful in promoting staff interaction and a sense of community.

The office floor plates suit a variety of layouts, from cellular offices to open plan, and this flexibility has encouraged tenants to break from traditional partitioned floors that are the predominant form in most local commercial buildings to a more informal, yet efficient workplace environment.

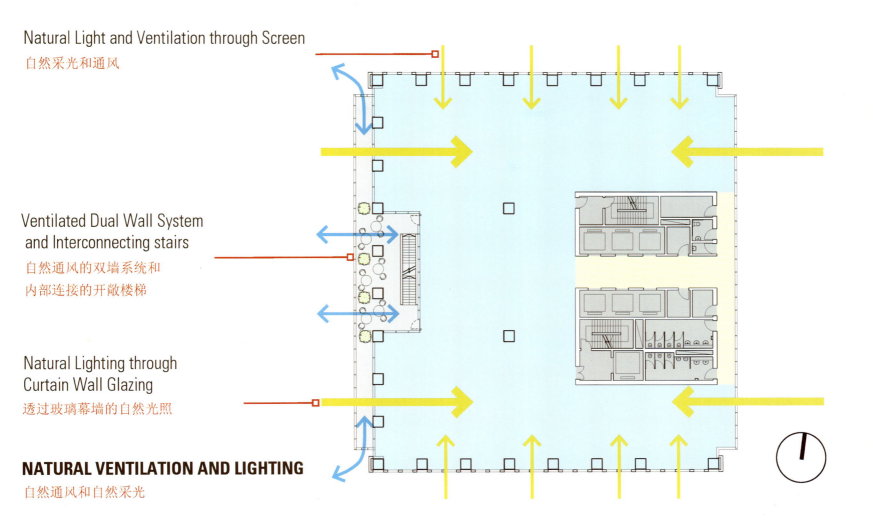

Natural Light and Ventilation through Screen
自然采光和通风

Ventilated Dual Wall System and Interconnecting stairs
自然通风的双墙系统和
内部连接的开敞楼梯

Natural Lighting through Curtain Wall Glazing
透过玻璃幕墙的自然光照

NATURAL VENTILATION AND LIGHTING
自然通风和自然采光

这栋建筑代表了该地区与众不同个性鲜明的建筑形式。两个细长的抛光花岗岩面板支撑着裙房上方的玻璃塔楼，也构成了大楼的主入口。许多 1m × 1m 的镶嵌式规则窗格都打开着，保证了自然通风，使大楼具有独特且模糊的比例感。由于规划限制，大楼只有 100 米高，而这些窗格弥补了大楼的不足。

在花岗岩面板的近处有着精致清晰的窗格。传统中国建筑运用窗格来框景和引入阳光，更重要的是，能够在保持与外界联系的同时创造出私密的感觉。中国招商银行的花岗岩框架既注重了对周围城区的影响，又充分考虑了人体比例，营造出银行建筑持久、隐秘之感，这对树立银行建筑的形象来说十分重要。

悬空玻璃建筑的西立面是低辐射双层玻璃系统，立面内部设有一系列相互连接的楼梯和大面积的温室花园，用来作为非正式会议空间。这样的流通空间在中国其他的商业大厦中并不常见，但是事实证明这些空间的使用非常成功，既利于员工互动，培养集体感，又改变了传统死板的工作环境。

双层墙面系统使建筑区域内大范围自然通风，显著减少了大楼的能量消耗。

个体使用者也可以调控大厦的大范围自然通风，而高效双层墙面和裙房屋顶可以收集雨水进行再利用。中国招商银行大厦的设计是对挑战性气候、功能性要求和经济预算做出的灵敏的回应。

重要的是，这栋大厦通过一系列的设计和功能创新，为社会的可持续发展也做出了巨大的贡献，如相互连接的楼梯和休息空间，在增强员工互动和集体感方面取得了成功。

办公楼的平面布局从单元间办公室到开放式布局，变化多样。这种空间的灵活性鼓励租户打破当地传统商业建筑中分隔平面的主要形式，创造出更加随意却高效的办公环境。

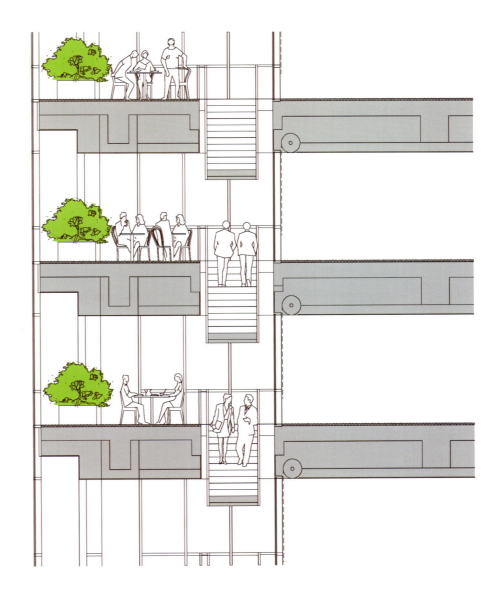

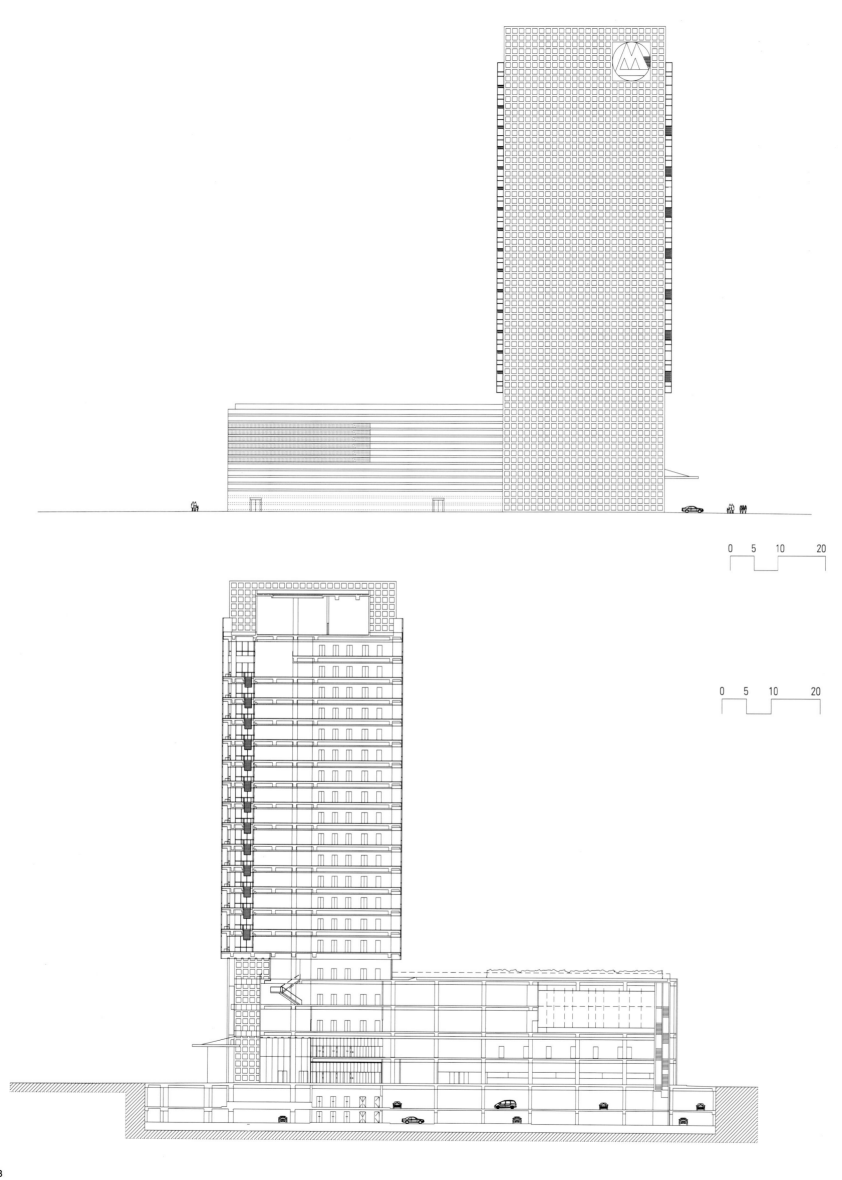

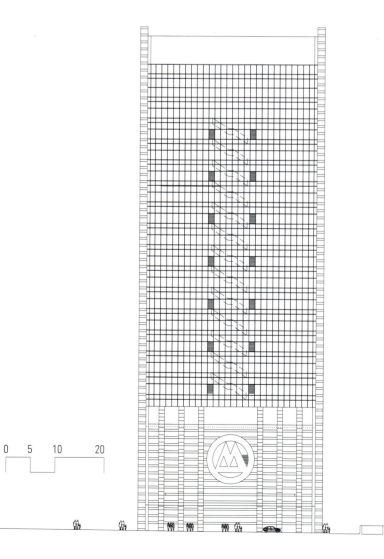

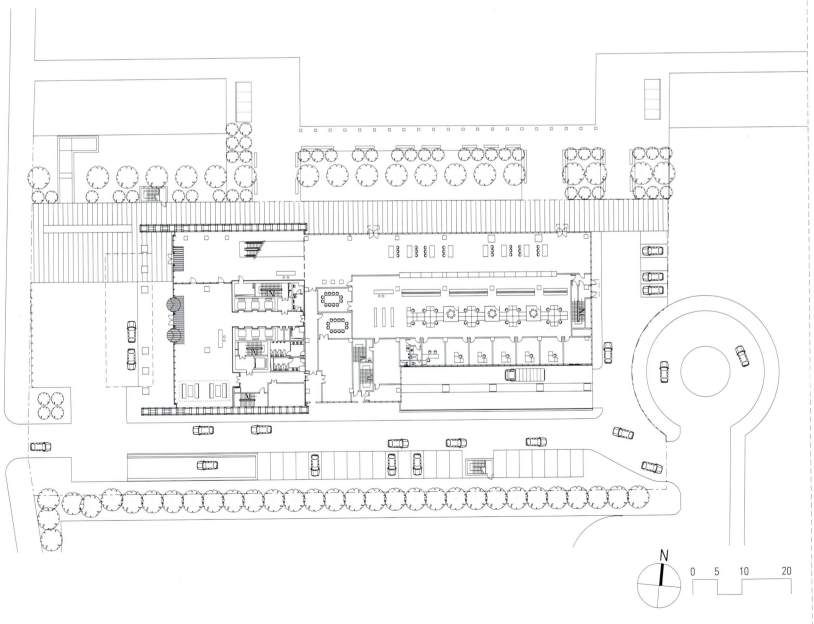

HONGZHU HOUSING SALES CENTER

鸿筑吾江接待中心

ARCHITECT
lab Modus

LOCATION
Taoyuan County, Taiwan, China

AREA
1,200 m²

PHOTOGRAPHER
Chih-Ming Wu

SUSTAINABLE & GREEN FEATURES 绿色特征

• *Use of recyclable materials* ／ 使用可回收材料

• *Special structural design to maximize daylight and natural ventilation, façade design to lighten heat island effect* ／ 特殊结构设计以最大限度利用自然光和自然通风，特殊表皮设计以减轻热岛效应

With an outer shell inspired by a dragon's scales, the Hongzhu Housing Sales Center in Taiwan minimizes direct sun exposure while encouraging natural daylighting. The façade is clad in a series of perforated metal panels bent and peeled up to create gaps in the armor that let in light and keep things cool inside. Designed by Taipei City firm, Lab Modus, the sales center creates a striking image in its urban location.

The Hongzhu Housing Sales Center is located on a narrow parcel on a street by the Taoyuan County exit of national highway. The three-storey building features a parking garage and entrance on the bottom with two floors of office and meeting space above. Taking full advantage of the site's footprint, the building expands to its extents and the massing takes its form from the available space.

Lab Modus took inspiration from the dragon, which is a lucky animal and considered a symbol of success and luck. In particular the dragon's scales were reinterpreted in an architectural form to aid in the energy performance of the building. The building's inner layer is clad in glass and then covered with a double layer of perforated metal panels. Articulation and gaps in the outer shell create a gap to allow natural ventilation and minimize solar heat gain. The design also allows diffuse and indirect light to enter the building. A dragon's outer shell thus reduces the building's need for air conditioning and artificial light.

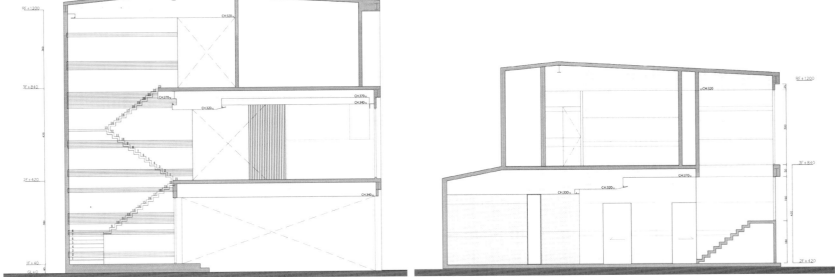

台湾鸿筑吾江接待中心的外形受到了龙鳞的启发，在减少阳光对建筑直射的同时增强了建筑的自然采光率。建筑外立面采用冲孔金属板，通过扭转打折，形成风洞，引入光线，使室内保持凉爽。鸿筑吾江接待中心由台北市 Lab Modus 建筑师事务所设计，这座销售中心的形象在其所在城市中非常引人注目。

鸿筑吾江接待中心位于桃源县，建在一块狭长的地皮上，邻近国道出口。这栋建筑高三层，停车场和入口在下层，上面的两层是办公室和会议室。该建筑充分利用其占地面积，在可用空间范围内进行最大程度的扩建。

龙是一种吉祥的动物，被视为成功与幸运的象征，Lab Modus 建筑师事务所从龙身上获得灵感。尤其是用建筑形式对龙鳞做了重新诠释，利于增强该建筑的能源效能。建筑的内层由玻璃包裹，又覆盖着双层冲孔金属板。金属结合部分在外墙上产生的风洞使建筑实现了自然通风，并减少了太阳辐射。这种设计同样可以使漫射光和间接光线进入建筑。因此，龙鳞形的外观减少了建筑对空调与人造光的需求。

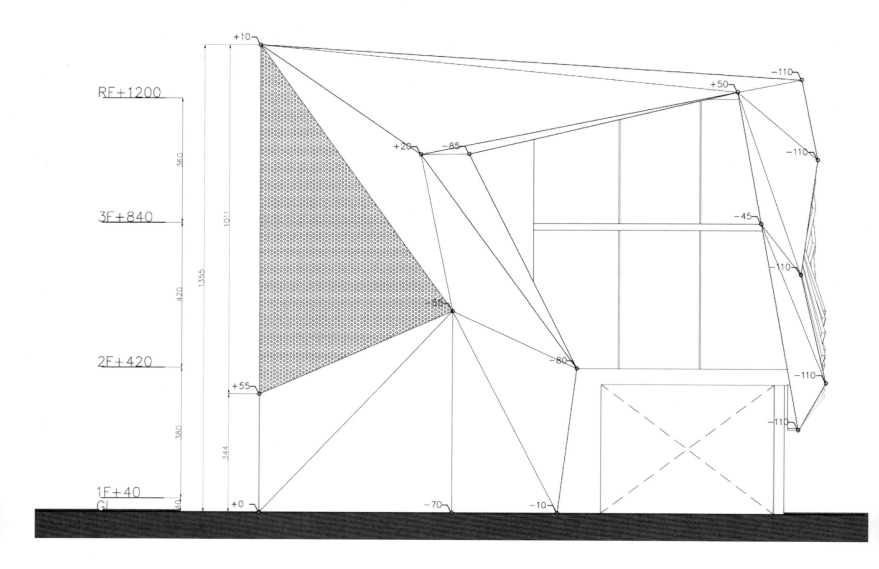

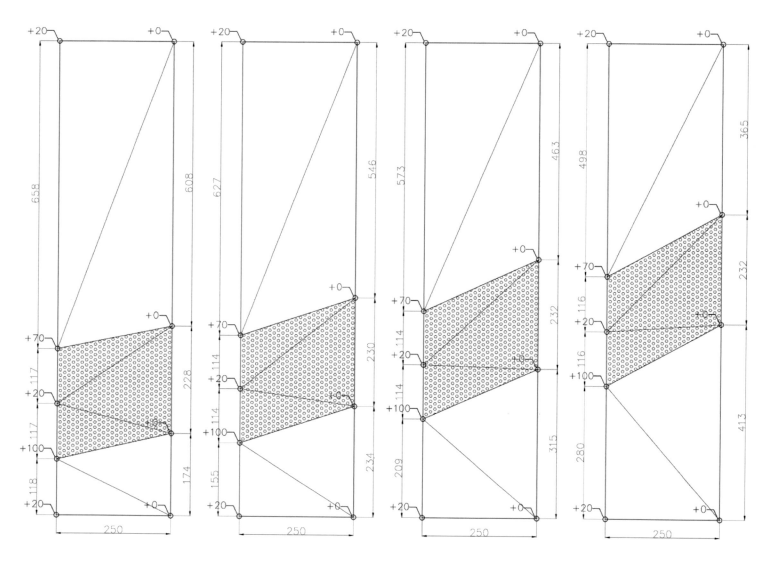

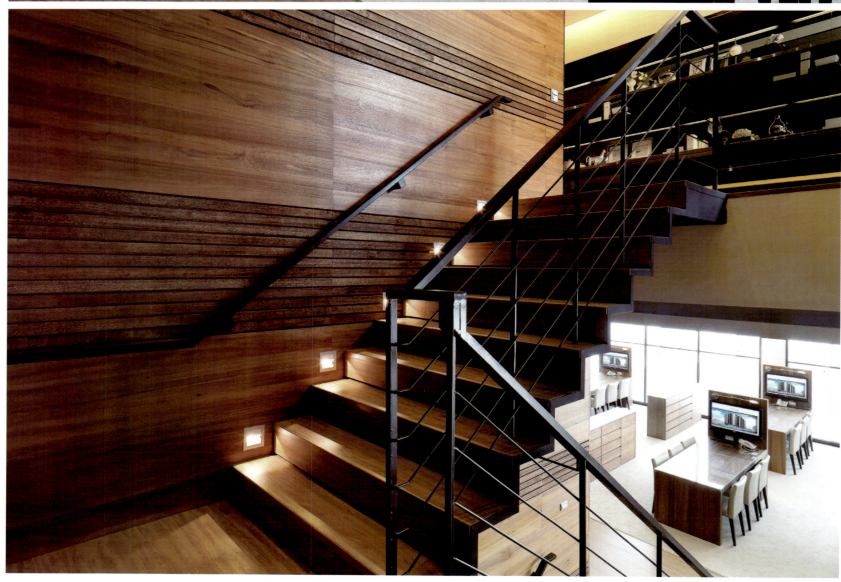

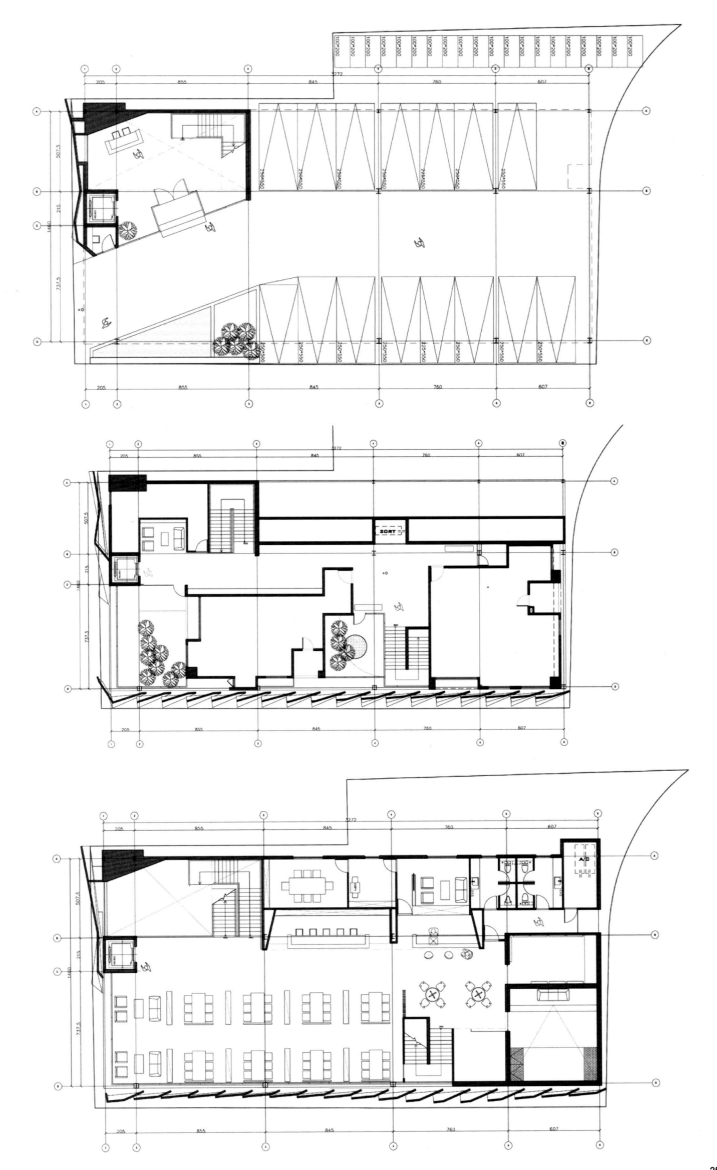

NUK COLLEGE OF HUMANITIES AND SOCIAL SCIENCES

高雄大学人文社会科学院

EEHW GOLD 台湾绿色建筑认证金奖认证

ARCHITECT
Malone Chang, Yu-lin Chen

FIRM
MAYU architects+

CLIENT
National Kaohsiung University

PROJECT TEAM
Malone CHANG & Yu-lin CHEN
(Architects), Kwantak AUYEUNG,
Ya-zhi KUO (Project Team), Yong-
chun CHANG, Ji-yang HUANG
(Construction Supervision)

STRUCTURAL ENGINEER
Envision Engineering Consultants

INTERIOR DESIGN
MAYU architects+

LANDSCAPE DESIGN
MAYU architects+

LOCATION
Kaohsiung, Taiwan, China

AREA
12,020 m²

PHOTOGRAPHER
Guei-Shiang Ke

SUSTAINABLE & GREEN FEATURES 绿色特征

• *Special window design to maximize sun shading and use of daylight and ventilation* ／ 特殊的窗户设计用以遮阳并引进自然光和自然通风

• *Highly efficient water reservoir to store water and reuse water* ／ 高效蓄水池用于蓄水和再利用水

• *Open foyer, courtyards, and arcade to maximize natural ventilation* ／ 开放的门厅、庭院和拱廊可保证充分利用自然通风

National Kaohsiung University is located in a floodplain where native species survive and flourish throughout the site. Therefore Eco-Campus is the goal of NUK, and half of its 82.5 acre (33.4 hectare) campus is preserved for ecological green area. The architectural strategy for this project is to create a college with strong bond to the site, and a building with various interstitial spaces that is suitable for tropical weather on the one hand and open to different uses on the other.

Under the extreme tropical climate, several measures are taken to ensure the sustainability of the project.

Sun – On the outer boundary of the architecture, punctured windows articulated with sun shading devices respond to the need to prevent incident direct sunlight. Vertical proportioned windows welcome reflected light and reduce the consumption of energy.

Water – The foundation water reservoir is installed in order to retain excess water and decrease runoffs caused by typhoons. The stored water is filtered and reused for irrigation.

Wind – The open foyer, courtyards, and arcade ensure unimpeded circulation of fresh air through major public spaces in the building. The natural ventilation reduces the reliance of air conditioning in milder seasons of spring and autumn.

Biological Diversity – In order to preserve the flourishing indigenous biological system, a central ecological pond is planned to provide diverse habitat for species. Endemic plants of Taiwan with different bloom season are chosen for new vegetation to further diversify the biological environment.

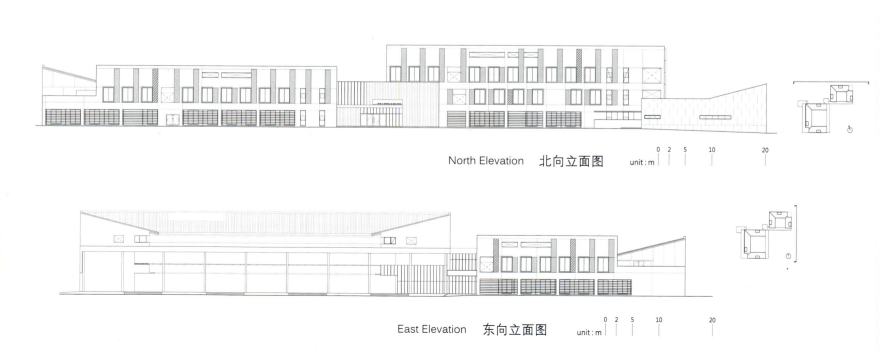

North Elevation 北向立面图 unit : m 0 2 5 10 20

East Elevation 东向立面图 unit : m 0 2 5 10 20

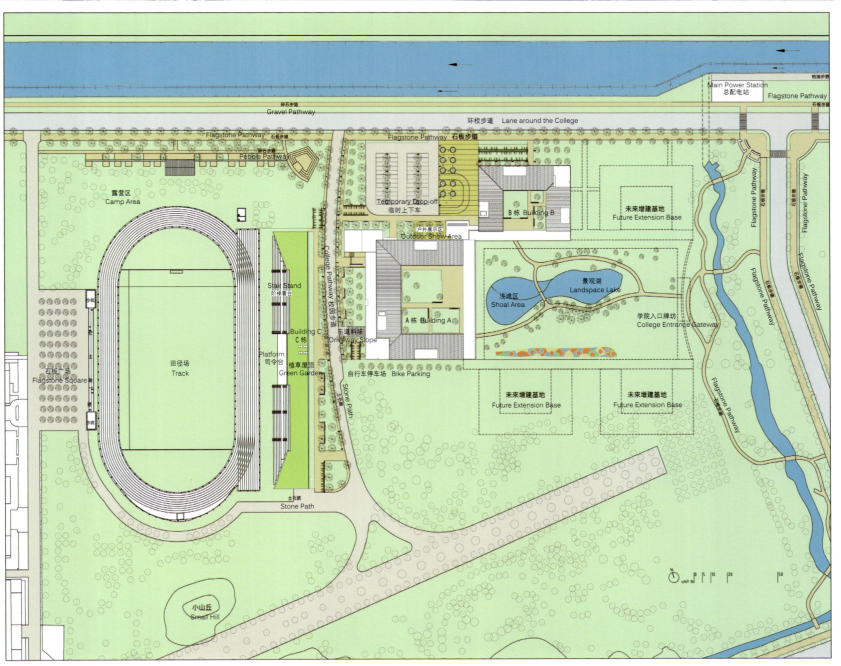

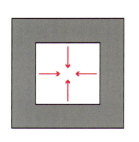

四合院
Quadrangle Courtyard

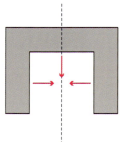

三合院
Three-section Courtyard

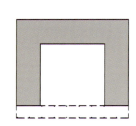

三点五合院
Three Point Five-section Courtyard

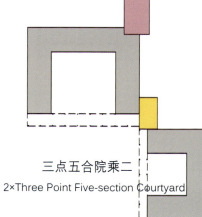

三点五合院乘二
2×Three Point Five-section Courtyard

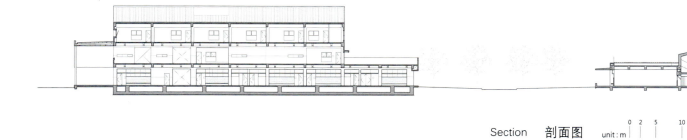

Section 剖面图 unit : m

0 2 5 10 20

Section 剖面图 unit : m

0 2 5 10 20

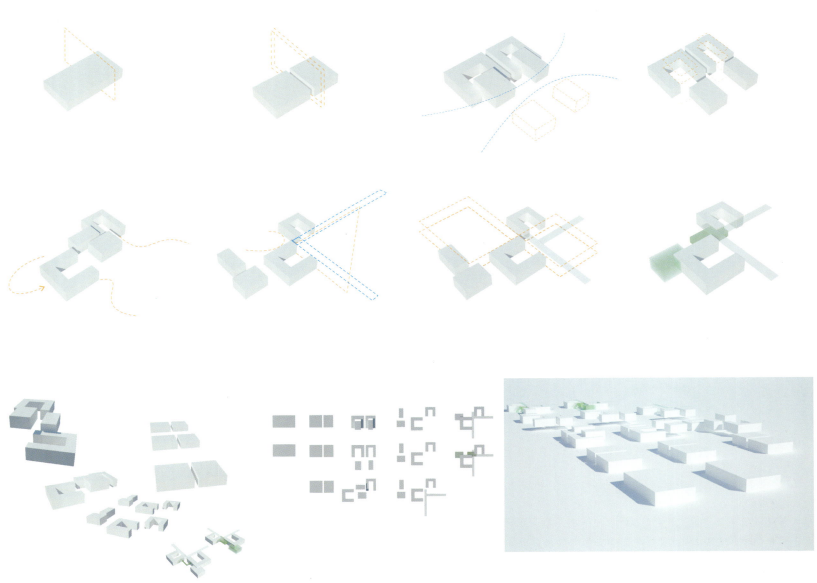

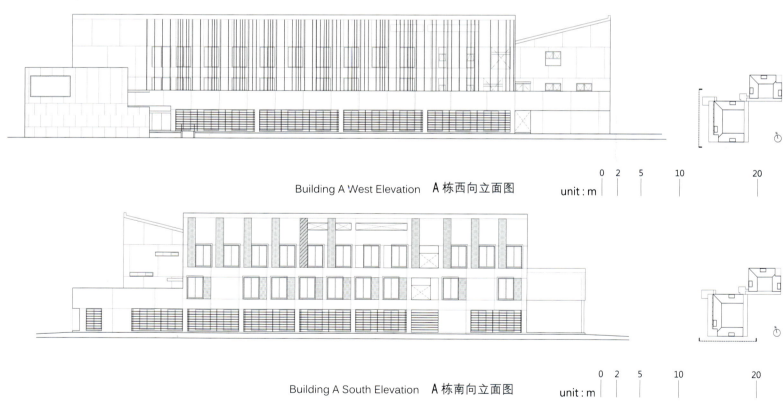

Building A West Elevation A栋西向立面图   unit : m   0 2 5   10   20

Building A South Elevation A栋南向立面图   unit : m   0 2 5   10   20

　　高雄大学（NUK）坐落在河漫滩上，本地物种在此繁衍生息。因此，NUK 的目标便是打造一个生态校园。学校占地 82.5 英亩（33.4 公顷），其中一半的面积打算用来建立生态绿地区。项目策略是要建立一所与周围环境紧密相连的大学。此外，所建大楼需有楼内间隔空间，使其既能适应热带气候，又可用于不同用途。

　　在极端热带气候的条件下，有关部门将采取多种措施保证该项目能够顺利进行。

　　阳光——在该建筑外缘，穿孔玻璃与遮阳设备一起，满足遮光之需。反射光从均匀垂直的窗户射进来，减少了能源的消耗。

　　水——安装喷水池是为了保留过量的水，以减少台风引起的径流。储存的水经过过滤，可用于灌溉。

　　风——开放式门厅、庭院及拱廊保证了新鲜空气在该建筑的主要公共空间内不断循环、畅通无阻；而自然通风则减少了建筑在春、秋等比较暖和的季节里对空调的依赖。

　　生物多样性——为了保护本土繁盛兴旺的生态系统，计划用中央生态水池为物种提供多样化的生存环境。此外，项目还选用了一些花期不同的台湾地区当地植物以培育新品种，从而使生态环境更加多样化。

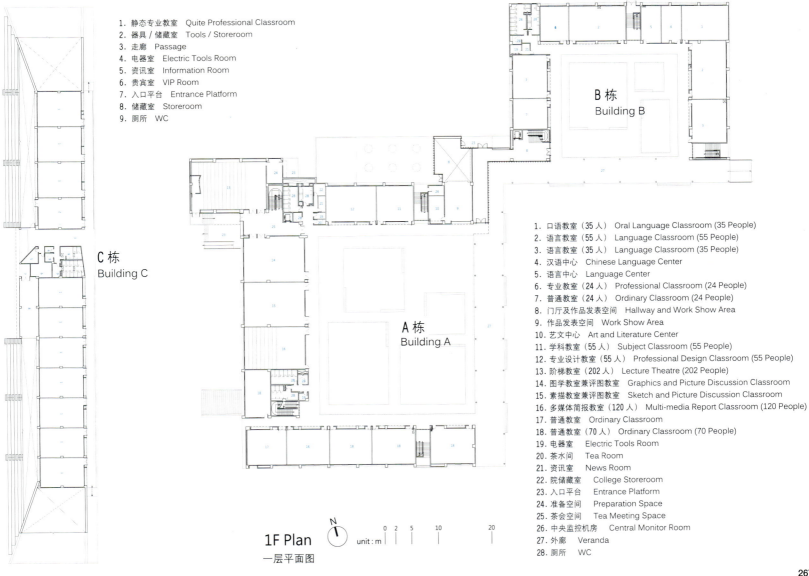

1. 静态专业教室　Quite Professional Classroom
2. 器具／储藏室　Tools / Storeroom
3. 走廊　Passage
4. 电器室　Electric Tools Room
5. 资讯室　Information Room
6. 贵宾室　VIP Room
7. 入口平台　Entrance Platform
8. 储藏室　Storeroom
9. 厕所　WC

C 栋
Building C

B 栋
Building B

A 栋
Building A

1. 口语教室（35 人）　Oral Language Classroom (35 People)
2. 语言教室（55 人）　Language Classroom (55 People)
3. 语言教室（35 人）　Language Classroom (35 People)
4. 汉语中心　Chinese Language Center
5. 语言中心　Language Center
6. 专业教室（24 人）　Professional Classroom (24 People)
7. 普通教室（24 人）　Ordinary Classroom (24 People)
8. 门厅及作品发表空间　Hallway and Work Show Area
9. 作品发表空间　Work Show Area
10. 艺文中心　Art and Literature Center
11. 学科教室（55 人）　Subject Classroom (55 People)
12. 专业设计教室（55 人）　Professional Design Classroom (55 People)
13. 阶梯教室（202 人）　Lecture Theatre (202 People)
14. 图学教室兼评图教室　Graphics and Picture Discussion Classroom
15. 素描教室兼评图教室　Sketch and Picture Discussion Classroom
16. 多媒体简报教室（120 人）　Multi-media Report Classroom (120 People)
17. 普通教室　Ordinary Classroom
18. 普通教室（70 人）　Ordinary Classroom (70 People)
19. 电器室　Electric Tools Room
20. 茶水间　Tea Room
21. 资讯室　News Room
22. 院储藏室　College Storeroom
23. 入口平台　Entrance Platform
24. 准备空间　Preparation Space
25. 茶会空间　Tea Meeting Space
26. 中央监控机房　Central Monitor Room
27. 外廊　Veranda
28. 厕所　WC

1F Plan
一层平面图

N

unit : m　0 2 5　10　20

267

ACADEMIC 3, CITY UNIVERSITY OF HONG KONG

香港城市大学学术楼（三）

BEAM PLATINUM (PROVISIONAL RATING) 香港 BEAM 绿色建筑认证铂金级认证（临时等级）
2012 GREEN BUILDING AWARD MERIT AWARD 2012 年度绿色建筑优秀奖

ARCHITECT
Ronald Lu & Partners

CLIENT
City University of Hong Kong

LOCATION
Hong Kong, China

AREA
37,300 m²

SUSTAINABLE & GREEN FEATURES 绿色特征

- *Recycling of rainwater for irrigation and cooling tower bleed-off water for flushing, recycling of A/C condense water for A/C make-up water, recycling waste* ／ 雨水回收用于灌溉，冷却塔流下的水用于冲洗，空调浓缩水回收后用作空调的补给水，废物回收

- *Use of Photovoltaic system, self-contained type PV cell landscape lights* ／ 使用光伏系统和带自供给型光伏电池的景观照明

- *Use of environmental friendly materials* ／ 使用环境友好型材料

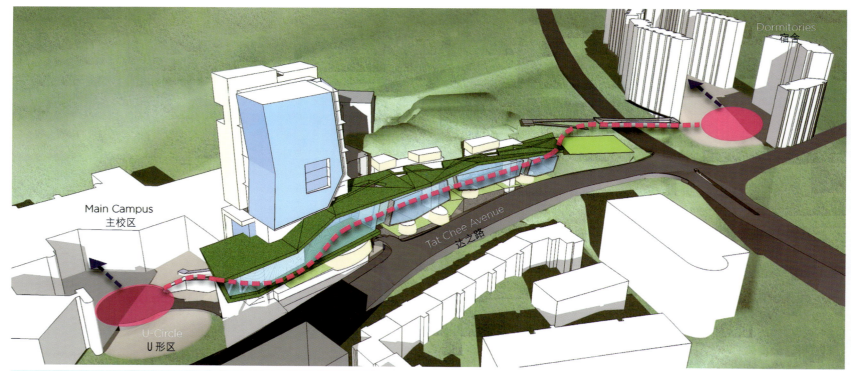

剖面图 Section

In order to solve the problem of an increasing number of students after the implementation of the 3/3/4 new standard, City University of Hong Kong has established a new teaching and administration building next to the existing University Square on Tat Chee Avenue to coordinate with future development of the university. Due to the project location and the height limitations, Ronald Lu & Partners is in conformity with four principles when designing the building: maintaining green environment of the site, increasing circulation space, constructing a green landmark building that is consistent with the background of Sierra Leone in sustainable development, and elevating the building to enhance the ground penetration and keep natural environment.

The new building is composed of high and low terraces, with Cornwall Street and students' dormitory to its north and the main campus to its south. The green terrace becomes the connecting passage, integrating the whole campus. Described as "Forest of Wisdom", it offers facilities for daily gathering and repose to teachers and students. Below the terrace, there are classrooms and lecture hall that can accommodate 600 people. By elevating the terrace above the ground, this design can allow the public to go through the building to the park behind. At the same time, natural elements can be ushered into the bottom of the building, inviting the public to go closer to nature.

The 12-storey building above the terrace is the highest building among the surrounding community. It is divided into two parts in visual effect in order to lower its sense of dimension. The design of the building separates different categories of departments and facilities according to the plan. The teaching and research laboratory are on the lower floors while administration offices are on the higher floors. The administration building sets up administrative departments such as board of directors of the university and the principal office. Besides, it also includes the activity room and two hanging gardens that can overlook Kowloon and Tat Chee Avenue.

Apart from offering various functional facilities to students and the staff, the new building will also become the new landmark of the university, providing more green space for the area and greatly improving the public facilities and circulation space of the campus.

3/3/4 新学季推行后，学生人数增多，为此香港城市大学于现有的达之路大学广场旁边，兴建了一座新的教学及行政大楼，以配合大学未来的发展。由于项目地点的规范及现有高度的限制，吕元祥建筑师事务所在设计大楼时秉持四个原则：维持地点绿化环境不减、增加流通空间、以可持续发展方向兴建一座与塞拉利昂背景相呼应的绿色地标性建筑、升高建筑以增加地面风透度及保持自然环境。

新大楼由高座及低座组成，北接歌和老街及大学学生宿舍，南通大学主校园，绿化的平台成为连接通道，整合了香港城市大学的校园。被喻为具有"智慧森林"的绿化平台，提供了师生日常聚会及憩息所用的设施，平台之下则为教室及可容纳

600 人的演讲厅等设备。平台高于地面的建筑设计，可以让公众穿过大楼到达后方的公园，也能让自然元素渗入楼底，为公众营造接触大自然的机会。

在平台之上的高 12 层的大楼是邻近小区内最高的建筑，于视觉效果上分为两部份，以减低其体量感。大楼的设计上有规划地将不同范畴的部门及设施分开，教学及研究实验室在低层，而行政办公室则位于高层。行政大楼设有大学董事会及校长办公室等大学行政部门，并包括两个可俯瞰九龙及达之路的空中花园及活动室。

新大楼除了为学生及职员提供各种功能设施外，亦将成为城市大学的新地标，为区域内提供更多绿化空间，还将大大地改善大学校园的公众设施及流通空间。

EAST ELEVATION
东立面图

The Rock

Twin Tower

Sky Court

Sky Lounge

The Park

Green Deck

The Path

University Street

Fleet of Knowledge

Lecture Theatre

Objects of Amenities

Podium

University Piazza

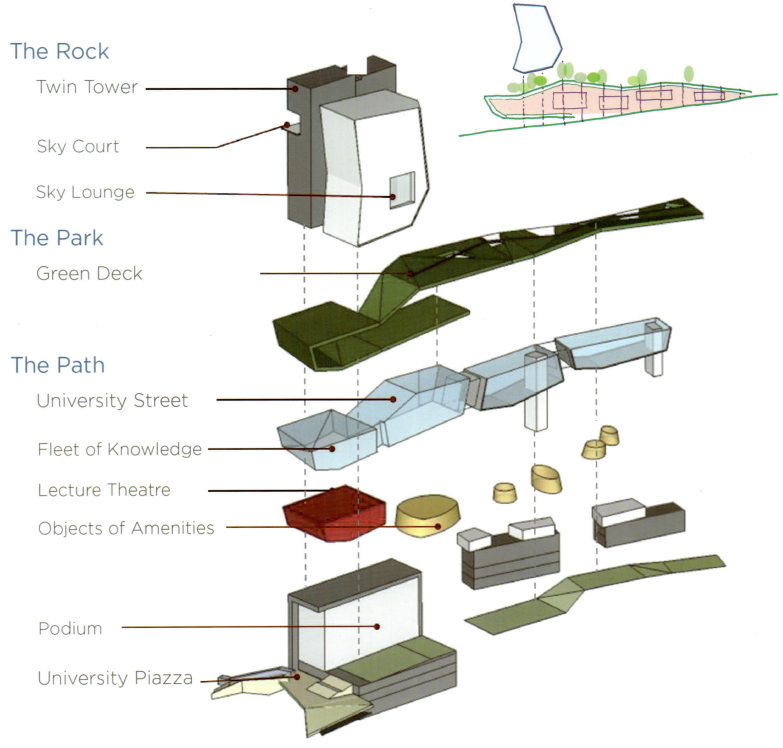

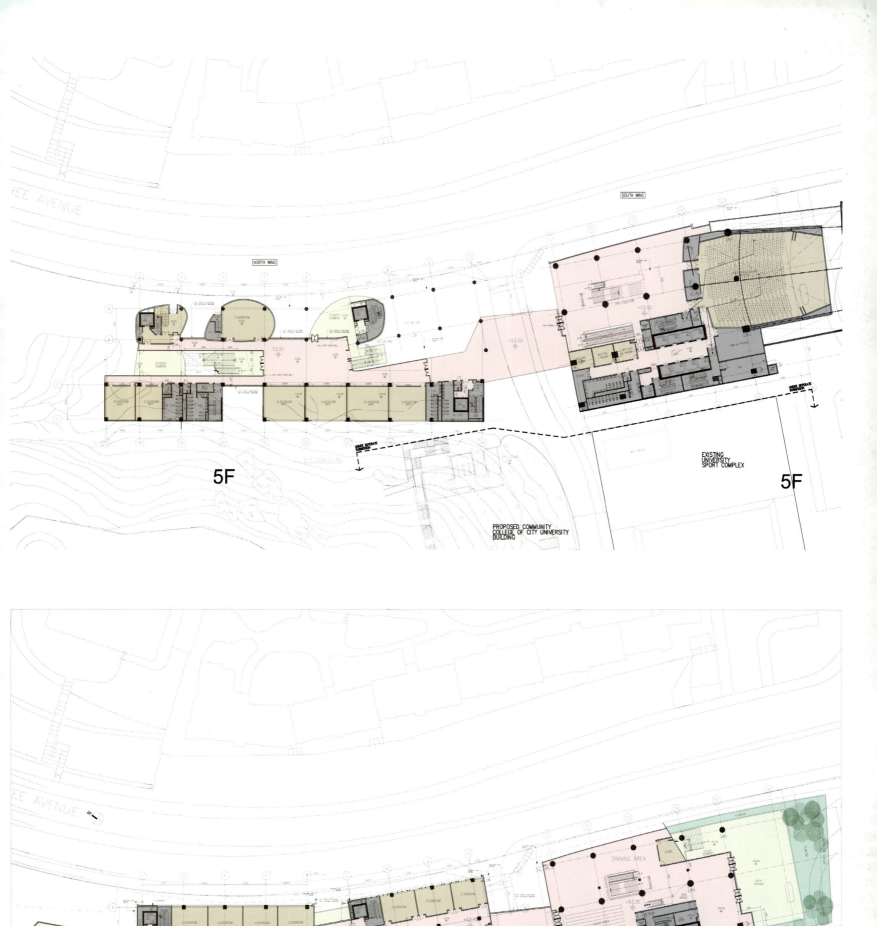

5F

NORTH WING

SOUTH WING

+53.00

+53.00

EXISTING
UNIVERSITY
SPORT COMPLEX

5F

PROPOSED COMMUNITY
COLLEGE OF CITY UNIVERSITY
BUILDING

EE AVENUE

DINNING AREA

+63.30

+63.00

7F

11th FLOOR PLAN
十一层平面图

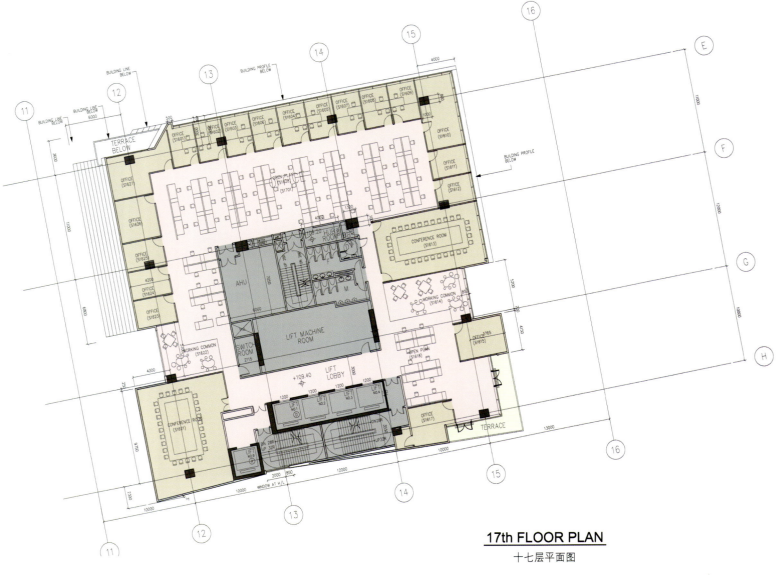

17th FLOOR PLAN

十七层平面图

SUZHOU HONG KONG AND CHINA GAS COMPANY TOWER

苏州港华燃气研发大楼

CHINA GREEN BUILDING 2 STAR CERTIFIED　中国绿色建筑二星级认证

ARCHITECT
Ming Lai Architects Inc.

LOCATION
Suzhou, Jiangsu Province, China

AREA
12,386.6 ㎡ (Floor Area), 74,970.7 ㎡ (Built Area)

SUSTAINABLE & GREEN FEATURES　绿色特征

- *Use of solar hot water system, energy-saving lighting system, and ventilation and heat recovery system* ／ 使用太阳能热水系统、节能灯具和风热回收系统
- *Green roof to recycle rainwater and lighten heat island effect* ／ 绿色屋顶用于雨水回收，减轻热岛效应

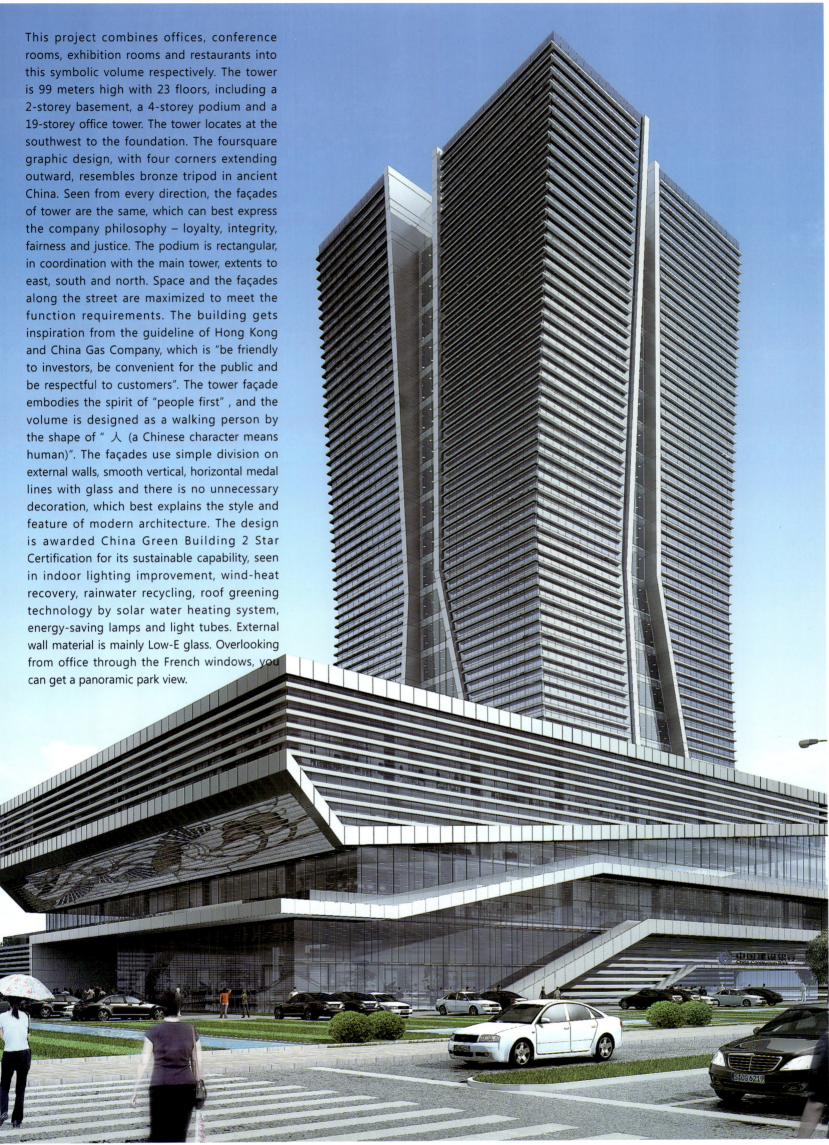

This project combines offices, conference rooms, exhibition rooms and restaurants into this symbolic volume respectively. The tower is 99 meters high with 23 floors, including a 2-storey basement, a 4-storey podium and a 19-storey office tower. The tower locates at the southwest to the foundation. The foursquare graphic design, with four corners extending outward, resembles bronze tripod in ancient China. Seen from every direction, the façades of tower are the same, which can best express the company philosophy – loyalty, integrity, fairness and justice. The podium is rectangular, in coordination with the main tower, extents to east, south and north. Space and the façades along the street are maximized to meet the function requirements. The building gets inspiration from the guideline of Hong Kong and China Gas Company, which is "be friendly to investors, be convenient for the public and be respectful to customers". The tower façade embodies the spirit of "people first" , and the volume is designed as a walking person by the shape of " 人 (a Chinese character means human)". The façades use simple division on external walls, smooth vertical, horizontal medal lines with glass and there is no unnecessary decoration, which best explains the style and feature of modern architecture. The design is awarded China Green Building 2 Star Certification for its sustainable capability, seen in indoor lighting improvement, wind-heat recovery, rainwater recycling, roof greening technology by solar water heating system, energy-saving lamps and light tubes. External wall material is mainly Low-E glass. Overlooking from office through the French windows, you can get a panoramic park view.

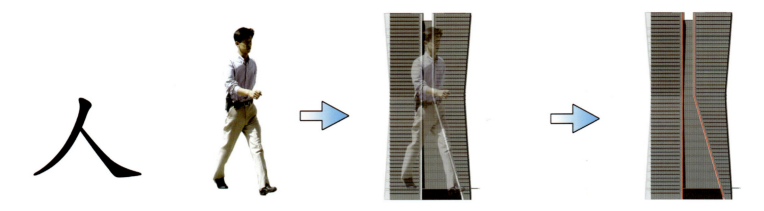

建筑意象的灵感来源于港华燃气公司的"亲商便民，以客为尊"的企业宗旨。
立面上将"以人为本"的精神融入其中，利用"人"字形将体量设计成一个人行走般的体态暗示，呈现标志性建筑的姿态。

The building gets inspiration from the guideline of Hong Kong and China Gas Company, which is "be friendly to investors, be convenient for the public and be respectful to customers".

The tower façade embodies the spirit of "people first", and the volume is designed as a walking person by the shape of "人 (a Chinese character means human)".

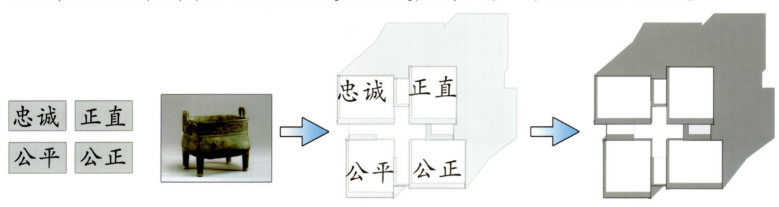

正方形平面上强调四个角落，与其内涵文化结合，分别代表忠诚、正直、公平、公正的企业理念。

The square façade underlines four corners, which can best express the company philosophy – loyalty, integrity, fairness and justice.

主建筑设计意向
DESIGN CONCEPT

技术经济指标： （实际面积以勘测放线数据为准）

一．规划建设净地面积：	12386.6m²
二．规划总建筑面积：	74803.1m²
（一）地上建筑面积(不含架空层建筑面积)：	53866.0m²
1 甲级写字楼建筑面积：	34951.1m²
其中：低区办公面积：	18055.5m²
高区办公面积：	16895.6m²
2 裙楼商业配套用房建筑面积：（1~4层）	18247.7m²
3 机房建筑面积：	667.2m²
（二）架空层建筑面积（不计入容积率）：	0m²
（三）地下建筑面积：	20937.1m²
1 地下机动车库面积：	17070.5m²
2 非机动车库面积：	726.5m²
3 设备用房面积：	3140.1m²
地下建筑层数：	2层
三．容积率：	4.35
四．建筑密度：	33.9%
五．绿地率：	31.5%
六．机动停车数：	448个
其中：地下停车数：	418个
地面停车数：	30个
七．非机动停车位：	404个
八．建筑层数及高度：	23层 98.9m

1:600
0 5   15 20m

总平面
SITE PLAN

本项目将办公、会议、展厅及餐饮等功能分别结合在这个标志性的体量当中。建筑高度为99米，共23层。体量上由2层地下室、4层裙房、及19层办公塔楼组成。塔楼位于基地西南角，平面设计方正，四个转角向外延伸，暗喻中国古代铜鼎的造型。自各个方向仰视，塔楼外观皆相同，与公司文化内涵结合，凸显港华忠诚、正直、公平、公正的企业理念。裙房体量配合塔楼的造型呈矩形，主立面沿东、南、北三个方向配置，空间及沿街立面最大化，满足功能上的需求。建筑意象的灵感来源于港华燃气公司的"亲商便民，以客为尊"的企业宗旨。塔楼立面造型上将"以

人为本"的精神融入其中，利用"人"字形将体量设计成一个人行走般的体态暗示。建筑立面采用简洁的外墙分割，流畅的垂直体量，水平金属线条与玻璃的融合，没有其他过多的装饰造型，体现了现代建筑的风貌。建筑通过太阳能热水系统、节能灯具、灯管改善室内采光、风热回收、雨水回收利用、屋顶绿化等技术措施来体现其可持续发展性，并获得了中国绿色建筑二星级认证。外墙材料以低辐射镀膜中空玻璃（Low-E 玻璃）为主，由办公室的落地玻璃向外眺望，园区景观一览无余。

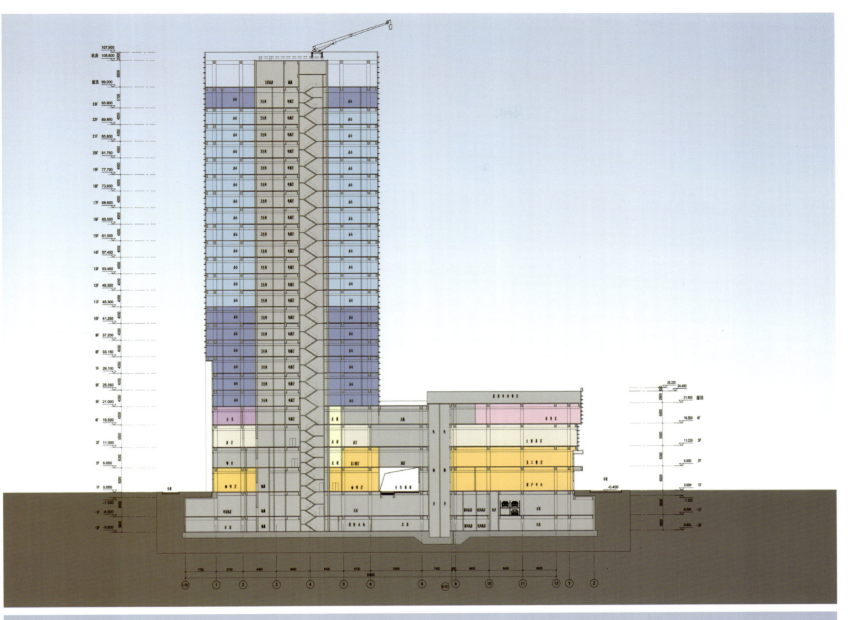

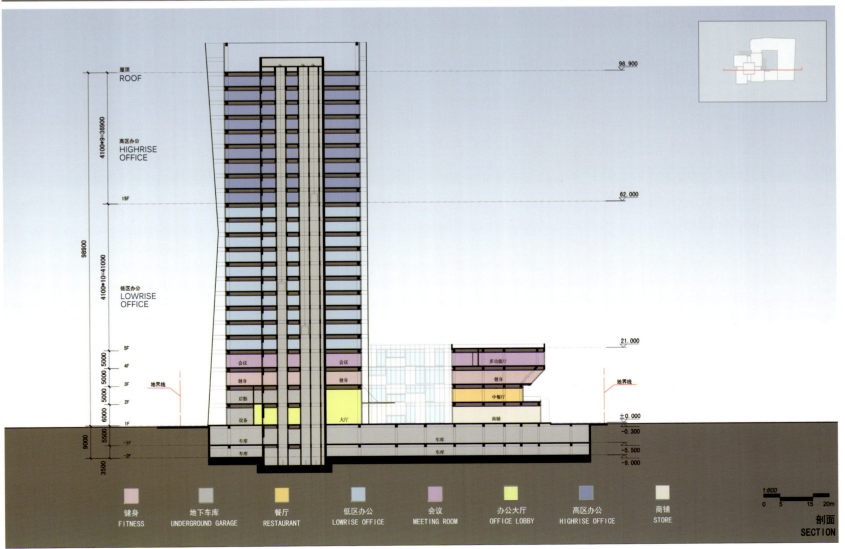

剖面
SECTION

健身　　　　地下车库　　　　　餐厅　　　　低区办公　　　　会议　　　　办公大厅　　　　高区办公　　　　商铺
FITNESS　　UNDERGROUND GARAGE　RESTAURANT　LOWRISE OFFICE　MEETING ROOM　OFFICE LOBBY　HIGHRISE OFFICE　STORE

剖面
SECTION

通讯系统	Telephone communication system

∠语音通讯 ∠Voice Transmission System
- 电信外线光缆与800对电信入户
- 电信机房设备容量10000门
- 语音增值功能齐全，语音邮箱，32方电话会议等各种电话功能

- Optical fiber connection and 800 pairs of UTP incoming cable to the building
- Telephone capacity :An immediate equipment capacity of 10,000 lines
- Comprehensive value added PABX service ,including voice message/mailbox，32 way telephone conferencing etc.

∠数据通讯 ∠Data Transmission System
- 外部光缆由中国电信区域骨干网关节点引入，传输带宽达到2.5G 并可随时扩容
- 齐全的接入方式：DDN、ISDN、FR、ATM等

- Direct connection to China Telecoms backbone network with an immediate bandwidth of 2.5G．Bandwidth can beadily expanded if necessary
- Various modes of connection available：DDN, ISDN, FR, ATM etc.

∠综合布线 ∠Structural Cabling
- 主干光线：传输带宽大于2.5G。双路由指向
- 重直主干线：语音采用三类大对数电缆，数据采用18芯多模光纤

- Optical Fiber backbone：A bandwidth of over 2.5Gbps, dual-routing direction
- Vertical Cat 3 UTP Cable for voice transmission and 18-strand muitl-model optical Fiber connected to PDS room at even floor

∠架空地板：办公楼层全面铺设钢制防静电架空网络线槽地板，高度10cm
∠无线局域网：无线传输速度11M
∠手机直放系统
∠有线电视系统

- 10cm high , anti-static electricity , steel raised flooring and trunking system Throughout the towers
- 11M Wireless LAN in pubic area
- Mobile phone signal amplification system
- Cable TV system

∠楼层网络机房 ∠Cable internet service room
- 机房专用空调系统：独立24小时空调冷却水供应
- 用电标准：1000W/㎡（双路市电）+后备电（柴油发电机）
- 数据传输：带宽2M以下由3类电缆提供，带宽2M以上由光缆提供
∠单独接地

- 24-hour chilled water supply
- Designed power capacity：1000W/㎡,back up power supply by emergency generator
- Data Transmission：Optical Fiber[bandwidth >2M] or CAT 3 Cable[bandwidth<2M]
- Power earthing and custom-designed professional earthing

∠电梯 ∠Elevator
∠客梯：8部，群控
高区：4部，额定速度：3.5米/秒
低区：4部，额定速度：3.0米/秒
额定载重：1350公斤
轿厢高度：2.5米
∠服务梯：2部，额定载重：1600公斤
∠额定等候时间：30秒

- 8 Passenger lifts, Destination Dispatching System
- 4 for High Zone Speed:3.5 meter per second
- 4 for Low Zone Speed:3.0 meter per second
- Each with a capacity of 1350kg
- Lift cabinet ceiling height:2.5m
- 2 Imported service lift for each tower , with a capacity of 1600kg
- Designed Waiting time: Max 30 seconds

∠强电系统 ∠Electrical system
∠电源：10KV双路进电供电
∠总用电负荷：10000KVA
∠办公区供电容量：110VA/平方米
∠后备发电机组：1600KVA

- Power supply：10KV, dual feed
- Total Power Capacity：10,000KVA
- Design power capacity for office：110VA/㎡
- Emergency power supply： 1,600KVA

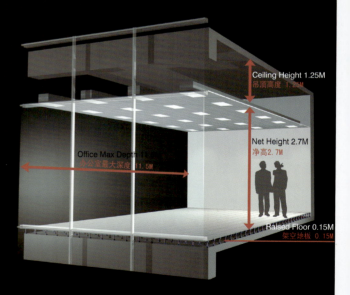

办公硬件系统建议
OFFICE TECHNICAL SPECIFICATION

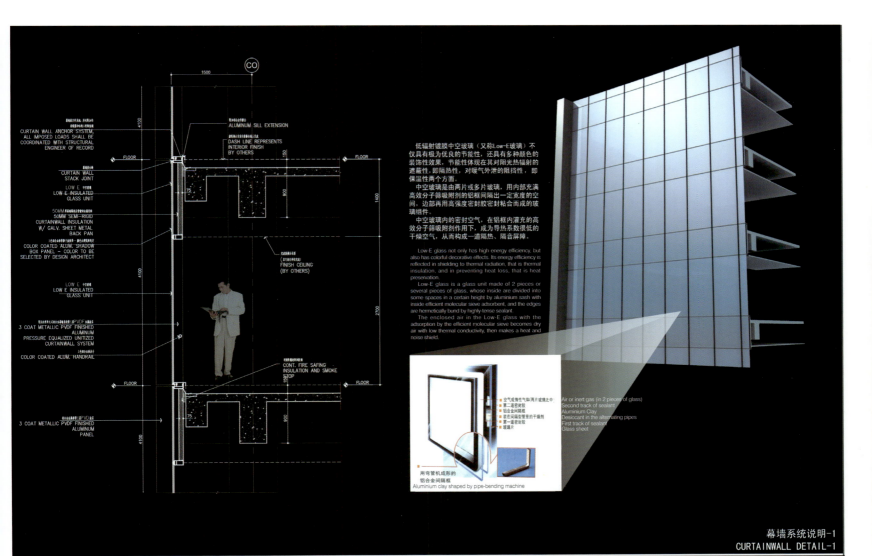

CURTAIN WALL ANCHOR SYSTEM, ALL IMPOSED LOADS SHALL BE COORDINATED WITH STRUCTURAL ENGINEER OF RECORD

CURTAIN WALL STACK JOINT

LOW E. LOW E INSULATED GLASS UNIT

50MM SEMI-RIGID CURTAINWALL INSULATION W/ GALV. SHEET METAL BACK PAN

COLOR COATED ALUM. SHADOW BOX PANEL - COLOR TO BE SELECTED BY DESIGN ARCHITECT

LOW E. LOW E INSULATED GLASS UNIT

3 COAT METALLIC PVDF FINISHED ALUMINUM PRESSURE EQUALIZED UNITIZED CURTAINWALL SYSTEM

COLOR COATED ALUM. HANDRAIL

3 COAT METALLIC PVDF FINISHED ALUMINUM PANEL

ALUMINUM SILL EXTENSION

DASH LINE REPRESENTS INTERIOR FINISH BY OTHERS

FINISH CEILING (BY OTHERS)

CONT. FIRE SAFING INSULATION AND SMOKE STOP

低辐射镀膜中空玻璃（又称Low-E玻璃）不仅具有极为优良的节能性，还具有多种颜色的装饰性效果，节能性体现在其对阳光热辐射的遮蔽性，即隔热性，对暖气外泄的阻挡性，即保温性两个方面。
中空玻璃是由两片或多片玻璃，用内部充满高效分子筛吸附剂的铝框间隔出一定宽度的空间，边部再用高强度密封胶密封粘合而成的玻璃组件。
中空玻璃内的密封空气，在铝框内灌充的高效分子筛吸附剂作用下，成为导热系数很低的干燥空气，从而构成一道隔热、隔音屏障。

Low-E glass not only has high energy efficiency, but also has colorful decorative effects. Its energy efficiency is reflected in shielding to thermal radiation, that is thermal insulation, and in preventing heat loss, that is heat preservation.
Low-E glass is a glass unit made of 2 pieces or several pieces of glass, whose inside are divided into some spaces in a certain height by aluminium sash with inside efficient molecular sieve adsorbent, and the edges are hermetically bund by highly-tense sealant.
The enclosed air in the Low-E glass with the adsorption by the efficient molecular sieve becomes dry air with low thermal conductivity, then makes a heat and noise shield.

空气或惰性气体(在两片玻璃之中) Air or inert gas (in 2 pieces of glass)
第二道密封胶 Second track of sealant
铝合金间隔框 Aluminium Clay
装在间隔型管里的干燥剂 Desiccant in the alternating pipes
第一道密封胶 First track of sealant
玻璃片 Glass sheet

用弯管机成形的铝合金间隔框
Aluminium clay shaped by pipe-bending machine

幕墙系统说明-1
CURTAINWALL DETAIL-1

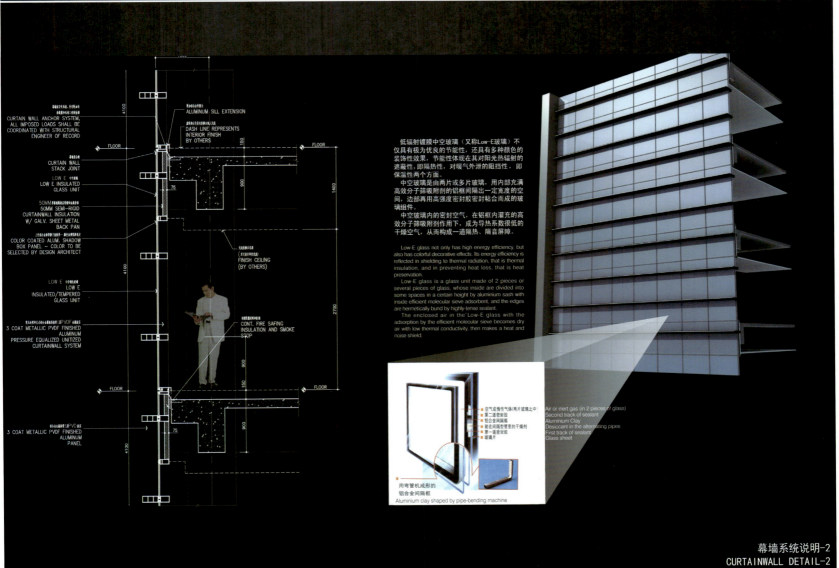

CURTAIN WALL ANCHOR SYSTEM, ALL IMPOSED LOADS SHALL BE COORDINATED WITH STRUCTURAL ENGINEER OF RECORD

CURTAIN WALL STACK JOINT

LOW E. LOW E INSULATED GLASS UNIT

50MM SEMI-RIGID CURTAINWALL INSULATION W/ GALV. SHEET METAL BACK PAN

COLOR COATED ALUM. SHADOW BOX PANEL - COLOR TO BE SELECTED BY DESIGN ARCHITECT

LOW E. LOW E INSULATED/TEMPERED GLASS UNIT

3 COAT METALLIC PVDF FINISHED ALUMINUM PRESSURE EQUALIZED UNITIZED CURTAINWALL SYSTEM

3 COAT METALLIC PVDF FINISHED ALUMINUM PANEL

ALUMINUM SILL EXTENSION

DASH LINE REPRESENTS INTERIOR FINISH BY OTHERS

FINISH CEILING (BY OTHERS)

CONT. FIRE SAFING INSULATION AND SMOKE STOP

低辐射镀膜中空玻璃（又称Low-E玻璃）不仅具有极为优良的节能性，还具有多种颜色的装饰性效果，节能性体现在其对阳光热辐射的遮蔽性，即隔热性，对暖气外泄的阻挡性，即保温性两个方面。
中空玻璃是由两片或多片玻璃，用内部充满高效分子筛吸附剂的铝框间隔出一定宽度的空间，边部再用高强度密封胶密封粘合而成的玻璃组件。
中空玻璃内的密封空气，在铝框内灌充的高效分子筛吸附剂作用下，成为导热系数很低的干燥空气，从而构成一道隔热、隔音屏障。

Low-E glass not only has high energy efficiency, but also has colorful decorative effects. Its energy efficiency is reflected in shielding to thermal radiation, that is thermal insulation, and in preventing heat loss, that is heat preservation.
Low-E glass is a glass unit made of 2 pieces or several pieces of glass, whose inside are divided into some spaces in a certain height by aluminium sash with inside efficient molecular sieve adsorbent, and the edges are hermetically bund by highly-tense sealant.
The enclosed air in the Low-E glass with the adsorption by the efficient molecular sieve becomes dry air with low thermal conductivity, then makes a heat and noise shield.

空气或惰性气体(在两片玻璃之中) Air or inert gas (in 2 pieces of glass)
第二道密封胶 Second track of sealant
铝合金间隔框 Aluminium Clay
装在间隔型管里的干燥剂 Desiccant in the alternating pipes
第一道密封胶 First track of sealant
玻璃片 Glass sheet

用弯管机成形的铝合金间隔框
Aluminium clay shaped by pipe-bending machine

幕墙系统说明-2
CURTAINWALL DETAIL-2

金 鸡 湖 大 道

道路红线
Road Red Line

退界线
Setback Line

自动银行
Self-Help-Bank

Company Product Show
港华产品展示

Bank
银行

Lobby
大厅

Lobby
大厅

Lobby
大厅

AC

E

Equipment
设备

Shop
商铺

Cafe
咖啡厅

Rear Service
后勤

Security Center
安保中心

Kitchen
厨房

黄 天 荡

通 园 路

商业电梯
RETAIL ELEVATOR

低区办公电梯
LOWRISE OFFICE ELEVATOR

高区办公电梯
HIGHRISE OFFICE ELEVATOR

办公服务电梯
OFFICE SERVICE ELEVATOR

车库客梯
GARAGE ELEVATOR

裙房货梯
PODIUM ELEVATOR

1:600
0   5   15 20m

N

底层平面
GROUND FLOOR PLAN

Kitchen
厨房

Western Restaurant
西餐厅

Equipment
设备

Storage
储藏

Lobby
大厅

Bottom
下空

AC

E

Chinese Restaurant
中餐厅

Rear Service
后勤

Staff Restaurant
员工餐厅

Kitchen
厨房

商业电梯
RETAIL ELEVATOR

低区办公电梯
LOWRISE OFFICE ELEVATOR

高区办公电梯
HIGHRISE OFFICE ELEVATOR

办公服务电梯
OFFICE SERVICE ELEVATOR

车库客梯
GARAGE ELEVATOR

裙房货梯
PODIUM ELEVATOR

1:500
0   5   10   20m

N

二层平面图
SECOND FLOOR PLAN

商业电梯
RETAIL
ELEVATOR

低区办公电梯
LOWRISE OFFICE
ELEVATOR

高区办公电梯
HIGHRISE OFFICE
ELEVATOR

办公服务电梯
OFFICE SERVICE
ELEVATOR

车库客梯
GARAGE
ELEVATOR

裙房货梯
PODIUM
ELEVATOR

1:500
0 5 10   20m

三层平面图
THIRD FLOOR PLAN

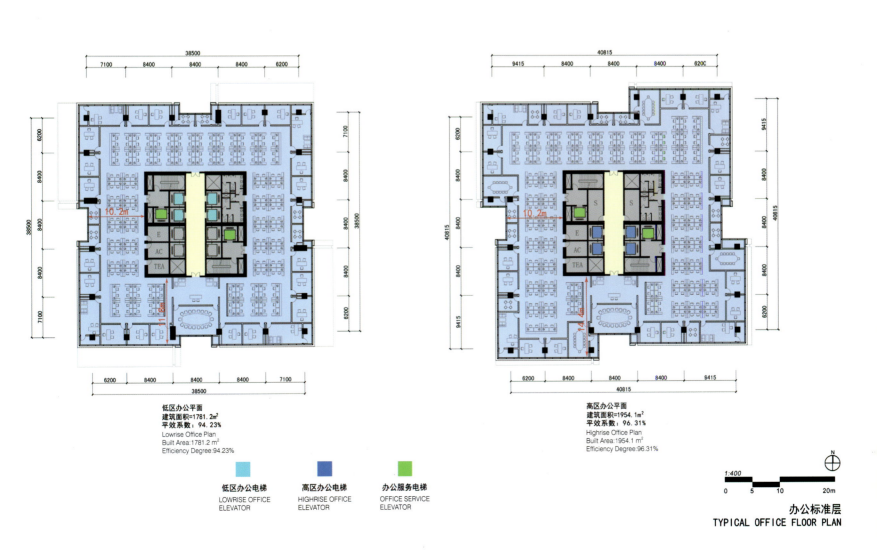

低区办公平面
建筑面积=1781.2m²
平效系数：94.23%
Lowrise Office Plan
Built Area:1781.2 m²
Efficiency Degree:94.23%

高区办公平面
建筑面积:1954.1m²
平效系数:96.31%
Highrise Office Plan
Built Area:1954.1 m²
Efficiency Degree:96.31%

低区办公电梯
LOWRISE OFFICE
ELEVATOR

高区办公电梯
HIGHRISE OFFICE
ELEVATOR

办公服务电梯
OFFICE SERVICE
ELEVATOR

1:400
0 5 10   20m

办公标准层
TYPICAL OFFICE FLOOR PLAN

GREEN ENERGY LABORATORY

绿色能源实验室

ARCHITECT
Laura Andreini, Marco Casamonti, Silvia Fabi, Giovanni Polazzi

FIRM
Archea Associati

PROJECT MANAGER
Enrico Ancilli

BUILDING SITE ASSISTANCE
Andrea Antonucci, Wang Xinfang

CLIENT
Jiao Tong University, Shanghai

LOCATION
Shanghai, China

AREA
1,500 m² (Plot Surface Area),
4,850 m² (Built Surface Area)

SUSTAINABLE & GREEN FEATURES　绿色特征

• *Access balconies to optimize energy consumption by accumulating heat in winter and aspirating the interior hot air in summer*／　可操作的窗户冬季蓄热，夏季排热，将节能最优化

• *Special building orientation and shape with the façade and glazed interior court to maximize natural ventilation and daylighting*／　特殊的建筑朝向和形状、建筑表皮和镶有玻璃的内院用于最大限度地利用自然通风和自然光

• *Highly energy-efficient double-skin façade and HVAC system*／　高效节能的双表皮系统和暖通空调系统

It was completed in April 2012 and opened on the following 19 May 2012 in the presence of the Minister of the Environment, Corrado Clini. Created as research center and laboratory for the analysis and diffusion of low carbon emission technologies in the construction and housing sector, the GEL is conceived as a compact body surrounding a central court, covered by a large skylight that can be opened or closed depending on the season, a solution chosen due to its functional characteristics in terms of distribution and energetic optimization.

The space, surrounded by access balconies, is configured as a void that optimizes energy consumption; on sunny winter days it functions as an accumulator of heat, and in summer it acts as a chimney, aspirating the hot air produced in the interior. The building has three floors with a total surface area of 1500 square meters above ground, and a maximum height of 20m. The first two floors host laboratories, meeting rooms, a control room, classrooms for the students and an exhibition space; every interior has windows on two sides, to the exterior and the inner court. The third floor hosts two sample apartments, the simulation of a two-room flat and a three-room flat covered by a pitched roof with photovoltaic panels, realized as platform for tests on residential types of spaces, to experiment with energy-efficient systems and buildings. The orientation of the building and its rectangular shape, along with the façade and the glazed interior court, are conceived to maximize the natural ventilation and to control exposure to the sun, in order to obtain an ideal interior climate with a minimum expenditure of energy. The façade, the distinctive feature of the exterior volume, consists of a double skin: an internal layer in glazed cells that provide waterproofing and insulation and an external one consisting of earthenware shutters that serve as sunscreens, to shade and regulate the illumination in the working spaces inside. The HVAC system has been designed on the basis of a main system (CHPC/WHP) combined with other, dedicated ones of smaller dimensions that are interchangeable according to the tests and research work done in the different laboratories.

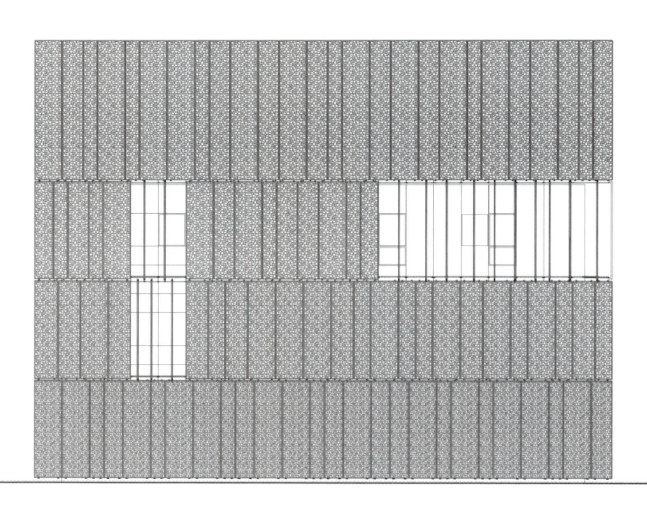

该项目于 2012 年 4 月完工。同年 5 月 19 日，环境部长科拉多·克里尼出席其落成典礼。绿色能源实验室既是一座研究中心，又是一个实验室，其主要用途是对低碳排放技术在建设与房地产行业的应用进行分析与传播。实验室结构紧凑，环绕在中央庭院周围。屋顶的大天窗可以根据季节变化打开或关闭，这一设计考虑到了其布局与能源优化等方面的功能特征。

这里四周围绕着阳台，形成了一个能够优化能源消耗的空间；在晴朗的冬日，它是一块蓄热电池，而到了夏季，它又是一个烟囱，排出室内产生的热气。该建筑共三层，地面总面积 1500 平方米，最大高度达 20 米。一、二楼配有实验室、会议室、控制室、学生教室以及展览区；每间房子的两侧都装有窗户，面朝室外与内院。三楼有两间样本室，分别进行两室公寓与三室公寓的模拟实验，房子的屋顶是倾斜的，上面配有光电板。这里是住宅类型空间的实验平台，对节能系统与建筑进行实验。建筑的朝向、矩形的形状，及其外观与镶有玻璃的天井都能在最大程度上增加自然通风，同时控制其暴露在阳光下的时间，从而以最少的能源消耗，获得理想的室内气候。建筑外观别具特色，由两层组成：内层镶有玻璃单元格，具有防水、保温的作用。外层由陶土遮板组成，起到遮光的作用，遮挡并调节内部工作区域的光照。空调通风系统的设计以主要系统（CHPC/WHP）为基础，同时与其他小规模专用系统相结合，它们可以根据不同实验室内的测试与研究工作而不断发生变化。

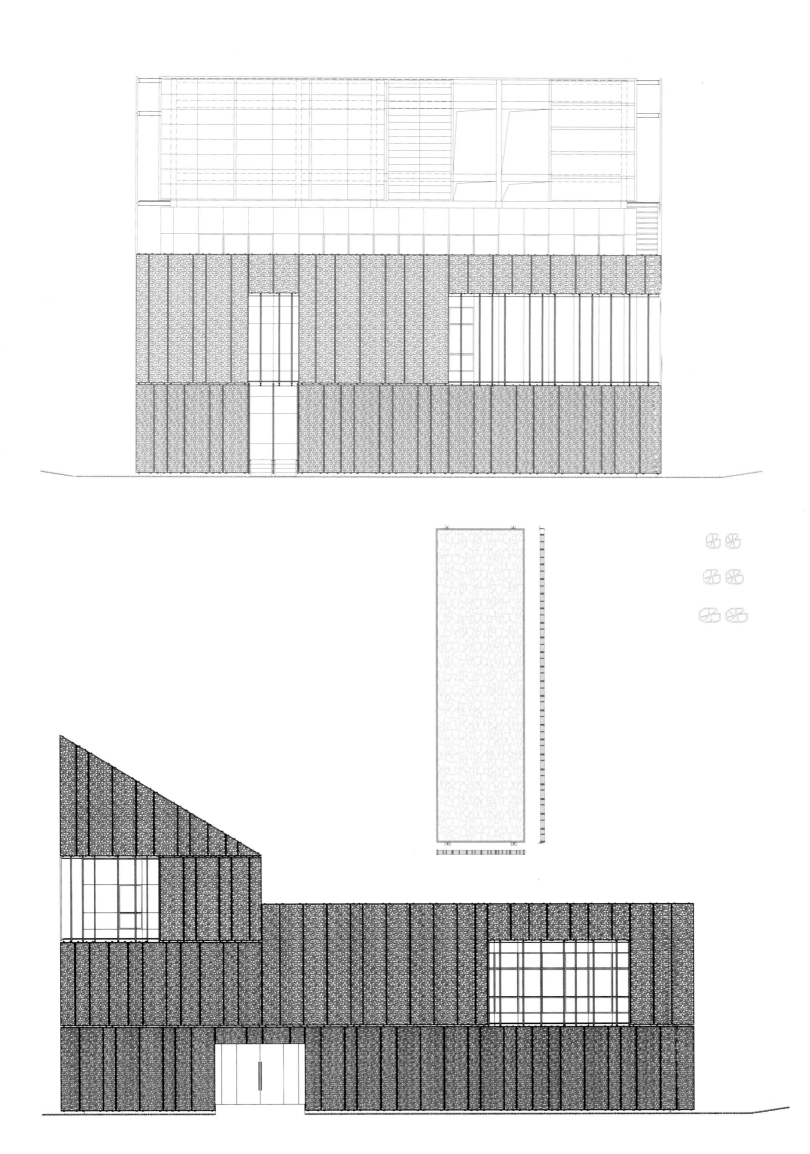

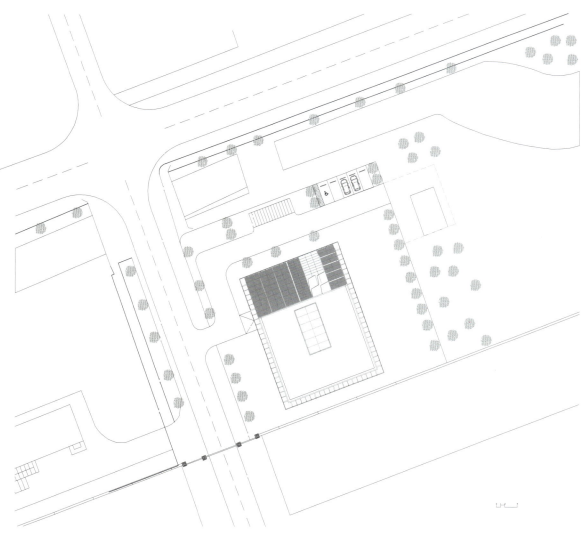

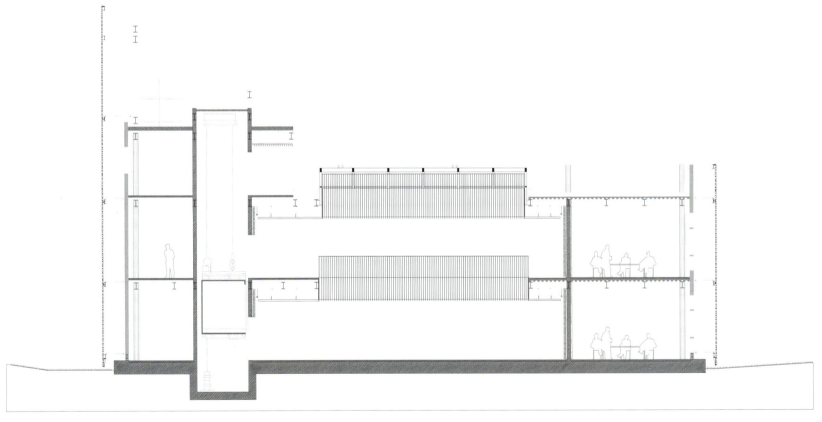

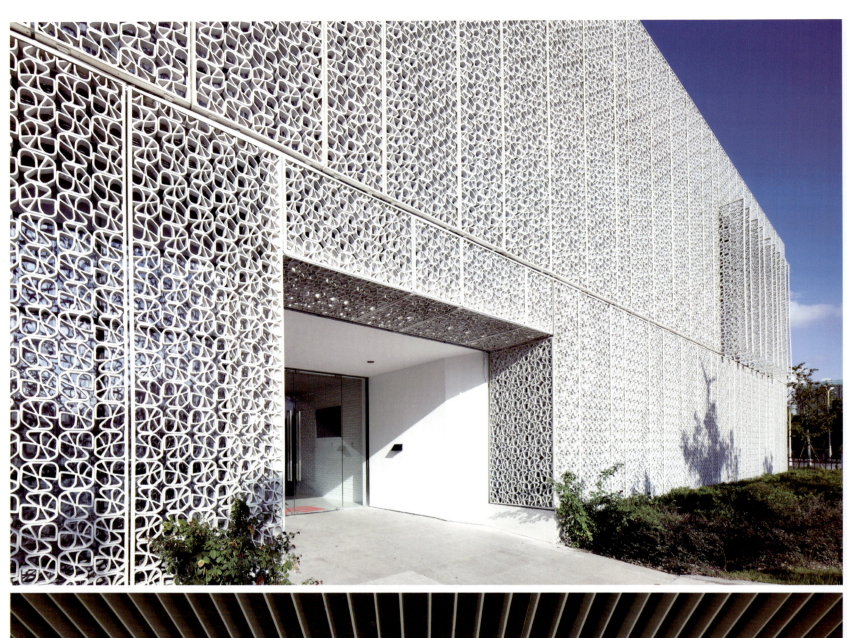

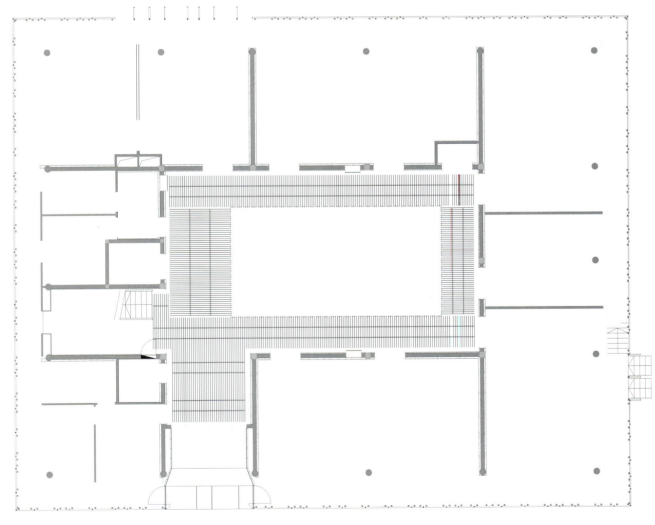

THE GUANGZHOU INTERNATIONAL FASHION CENTER

广州国际时尚中心

LEED-NC GOLD REGISTERED (ART GALLERY AND FLAGSHIP)
LEED NC & CS SILVER REGISTERED (OFFICE TOWERS)

LEED-NC 黄金级认证（艺术画廊和旗舰店）
LEED NC & CS 白银级认证（办公楼）

ARCHITECT

Jeffery Heller, Clark Manus

FIRM

Heller Manus Architects

LOCATION

Guangzhou, Guangdong Province, China

SUSTAINABLE & GREEN FEATURES 绿色特征

• *Operable windows to ensure natural ventilation and reduce use of air conditioning* ／ 可操作窗户确保自然通风，减少空调的使用

• *Use of local, rapidly renewable, recyclable materials* ／ 使用当地的、快速可再生的和可回收的材料

• *Energy efficient lighting systems, solar domestic hot water (DHW) system, low-flow plumbing systems, heat and energy recovery systems and HVAC system* ／ 高效节能的照明系统、太阳能生活热水系统、低流量管道系统、热量和能源回收系统和暖通空调系统

The Guangzhou International Fashion Center will house the new headquarters for Canudilo, a leading high-end men's clothing, accessories and luxury brand retailer in China, as well as substantial office and retail space, gallery for fashion exhibitions, conference center, roof gardens and underground parking. The Fashion Center will pursue an environmentally sustainable design plan with energy-saving and green building methods. The two high-rise commercial office towers are seeking LEED Silver certification, and the art gallery and flagship store are seeking LEED Gold certification.

Building Sustainability

• Operable windows where appropriate for thermal comfort;

• Use locally sourced materials;

• Use rapidly renewable materials;

• Use materials with high recycled content;

• Use materials with low VOCs, formaldehyde free to reduce indoor air pollution;

• Recycling program, green cleaning, HVAC performance monitoring to insure continued.

Sustainability Building MEP Sustainability & Energy Optimizing Strategies

• High performance glazing, fritting and exterior shading reduce heat gain in building;

• Building-integrated photovoltaic (BIPV) and roof-top PV systems;

• Utilize energy efficient lighting systems, especially in common areas; maximize daylight;

• Sub-metering individual lease spaces encourage tenant energy conservation;

• Solar domestic hot water (DHW) system;

• Low-flow plumbing systems reduce potable water use;

• Utilize high-efficiency machinery;

• Heat and energy recovery systems use waste energy to heat;

• Enhanced commissioning insures energy-related systems continue to perform to optimal standards;

• Zero use of CFC-based refrigerants in HVAC&R systems to reduce ozone depletion.

卡奴迪路是奢华零售品牌，在中国主要经营高端男装和饰品。该品牌将新总部落户在广州国际时尚中心。时尚中心内还设有大量办公区、商品零售区、时尚展览长廊、会议中心、屋顶花园和地下停车场。时尚中心追求环境可持续发展的设计理念，采用节能减排和绿色的建筑方法。两栋高耸的商务办公大楼正在竞争能源与环境设计先锋奖银奖，而艺术长廊和旗舰店则将参与能源与环境设计先锋奖金奖的争夺。

建筑可持续性
• 可操控窗户能够保持热舒适性；
• 使用当地资源材料；
• 使用迅速可再生的材料；
• 使用回收物质含量高的材料；
• 使用挥发性有机污染物含量低的、无甲醛的材料，减少室内空气污染；
• 循环系统、绿色清洁、暖通空调性能监测，以确保可持续性。

建筑 MEP 可持续和能源优化策略
• 高性能的玻璃、釉料和外部遮阳减少了建筑的热量吸收；
• 光伏建筑一体化和屋顶光伏系统；
• 使用节能照明系统，尤其是公共区域，将日光效果最大化；
• 辅助计量每个租赁空间，鼓励租户节约能源；
• 使用生活热水系统；
• 低流量管道系统减少饮用水的使用；
• 使用高效机械装置；
• 热量能源回收系统将废弃能源转化为热量；
• 强化性能验证确保了能源相关系统继续按照最佳标准运行；
• 空调系统采用不含氟制冷剂，以减轻对臭氧层的破坏。

品牌旗舰店
CANUDILO
FLAGSHIP STORE

A1塔楼
A1 TOWER

A2塔楼
A2 TOWER

艺术馆
ART GALLERY

Waterwall

Rockery

Reinforced
Turf Paving

Waterwall
Seat Walls

Connection to
Flower Mountain

Lantern

Reflecting
Pool

场地设计图例
Site Plan

1	水池	Reflecting Pool
2	假山	Rockery
3	强化绿地铺装	Reinforced Turf Paving
4	灯饰	Lantern
5	自然山体	Natural Hill
6	到达停车区	Passenger Drop-Off Area
7	地下车库入口	Parking Entrance
8	地下车库出口	Parking Exit
9	中心广场	Central Plaza
10	隔离绿带	Landscaping

首层平面图 Ground Level Plan
1:600

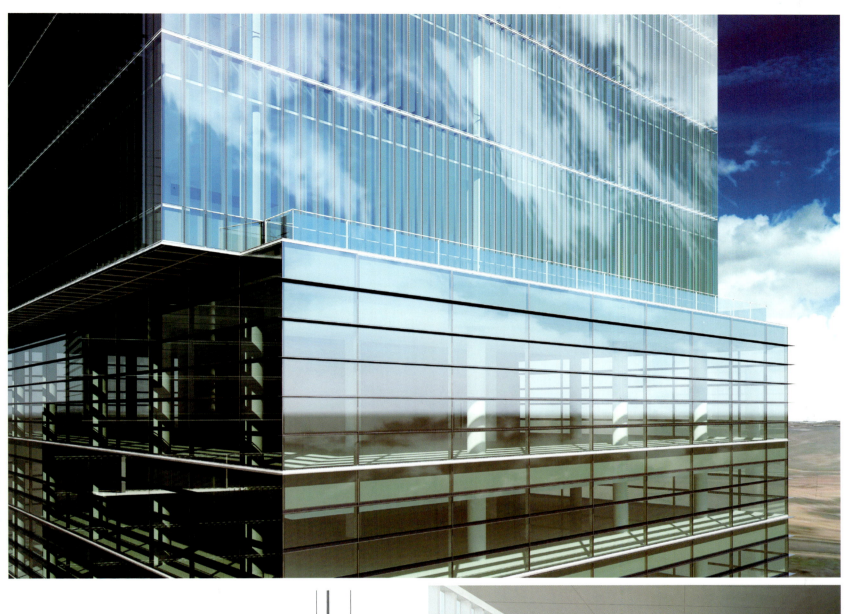

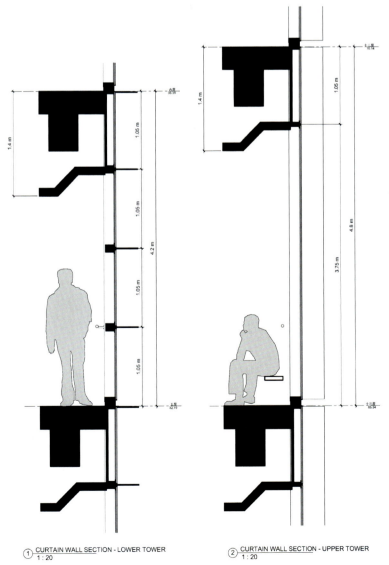

1 CURTAIN WALL SECTION - LOWER TOWER
1 : 20

2 CURTAIN WALL SECTION - UPPER TOWER
1 : 20

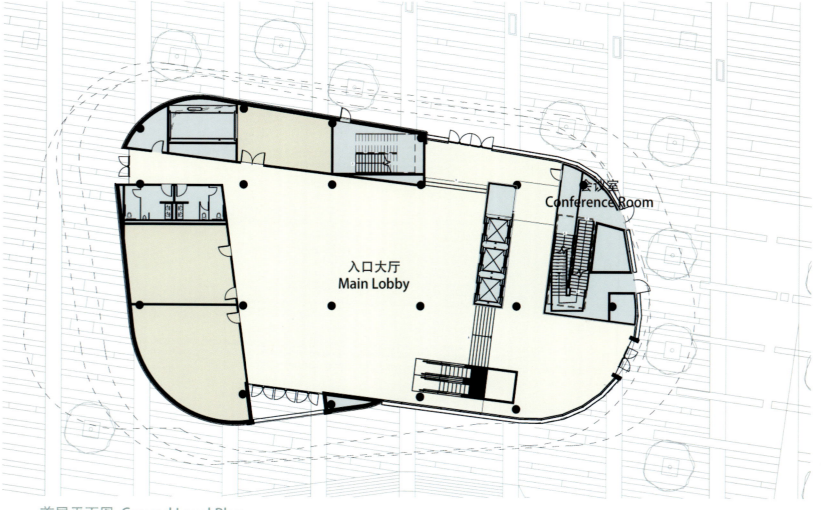

首层平面图 Ground Level Plan
1 : 400

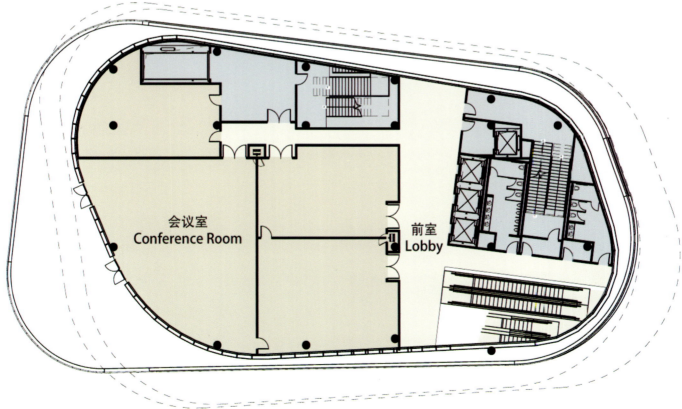

三层夹层平面 Level 03 Mezzanine Plan
1 : 400

GUODIAN NINGXIA SOLAR CO., LTD
OFFICE BUILDING

国电宁夏太阳能有限公司办公楼

ARCHITECT
BLVD INTERNATIONAL INC.

LOCATION
Shizuishan, the Ningxia Hui Autonomous Region, China

AREA
8,230 m²

PHOTOGRAPHER
Du Yun

SUSTAINABLE & GREEN FEATURES　绿色特征

• Use of energy saving façade material and solar energy ／ 使用节能的表皮材料和太阳能

One of the GUODIAN Ningxia Solar Co.,LTD's key products is Polycrystalline Silicon, a high-tech product widely used for the creation of semiconductors and solar batteries. The office building is a combination of the impressiveness of a large state-owned enterprise and high-technology. The architectural form is combined with the product feature of the enterprise, integrating the shape of poly silicon with inner space, utilizing the enterprise feature of solar energy conservation and environmental protection to introduce the ecological architecture concept for the building design.

Ningxia is located in the northwest inland area of China, an area in the temperate arid region, and is therefore, frequently visited by high wind and sandstorm. The façade of the building is made of alloy galvanized corrugated plates with different colors and unique shapes combined together, forming a strong modern sense. This kind of material is erosion-proof and the metal colors can best illustrate the high-tech quality of the modernized enterprise. The building is highlighted by bright red, standing out of the surrounding yellow-grey environment, and becomes the focus of the region. The west side of the façade mesh structure appears to be plain, but it is an extraordinary ecological forestation area. In spring and summer, the vertical planting can come into play, making the building harmonious with nature.

A dazzling Diamond – The beauty of the architecture lies not only in the construction itself, but also in its interaction with the surrounding. The office building, different from traditional buildings that have their direct entry ways, is surrounded by "green polylines" which overlap each other in shades of dark and light, resembling ripples of water created when a diamond is cast in. The location of the building in the Shizuishan area at the foot of Mount Helan also proved to be an inspiration with the design. A collection of selected little stones taken from this area are placed in a sequence creating an artistic impression of "withered and desolate". Such native plant species, as Chloris virgata Swartz, Berberis Thunbergii DC, Caryopteris Clandonensis Worcester Gold, Potentila Fruticosa, Rhus Typhina L and Hippophae Rhamnoides Linn are also used to produce seasonal variance of mixed colors and various combinations, creating a powerful oasis effect.

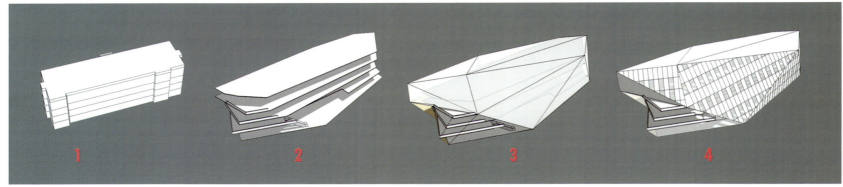

1   2   3   4

Features of architectural style and the process of the evolution

造型特点及演变过程

本建筑设计从建筑室内空间需要和造型出发，结合企业产品特点，将多晶硅矿石的造型和现代化的审美情趣结合，打造出别具一格的新型的建筑空间体验，利用太阳能节能环保的企业特性，将生态建筑的概念引入建筑设计当中。

The architectural form is combined with the product feature of the enterprise, integrating the shape of poly silicon with inner space, utilizing the enterprise feature of solar energy conservation and environmental protection to introduce the ecological architecture concept for the building design.

国电宁夏太阳能有限公司是以多晶硅为主要产品的企业。多晶硅的需求主要来自于半导体和太阳能电池，是高集成的高科技产品。因而从办公楼的建筑上我们不但能读出一个大型国企的气度，更能看到高科技的内涵。建筑外形充分结合了企业产品的特点，将多晶硅矿石的造型和室内空间需要相结合，同时利用了太阳能节能环保企业特性，将生态建筑的概念引入建筑设计当中。

宁夏身居西北内陆，地处中温带干旱区，大风、沙尘暴天气出现次数较多。因而建筑立面是由现代感很强的合金镀锌波纹板，以不同的颜色和独特的造型特点拼接而成的。这种材料不但能防止风沙腐蚀，更能让金属色彩烘托出现代化企业的高科技性质。鲜艳的红色，使建筑从周边灰黄的环境中跳出来，成为了此区域的视觉焦点。西侧的立面网状结构看似平淡无奇，其实是一块神奇的生态绿化的区域。在春夏季节，可以实现立体绿化的效果，使得建筑更加和谐，融入自然。

璀璨的钻石——建筑之美，不仅在于其自身的璀璨，更在于其与周边环境的互补。这个办公楼一改传统办公空间直达正入的方式，而是利用"绿意的折线"环绕在"钻石"周围，这些折线明暗相间、层层叠叠，仿佛钻石投在水中泛起的涟漪。石嘴山地区位于贺兰山脚下，因而设计就地取材把产自于贺兰山的小石头通过按一定顺序排放，形成"枯与荒"的艺术氛围。同时还运用了当地原生植物，包括虎尾草、柴叶小檗、金叶莸、金露梅、火炬树、沙棘等，不同的颜色相互交织，形态各异的植物组合，营造出了强大的绿洲效应。

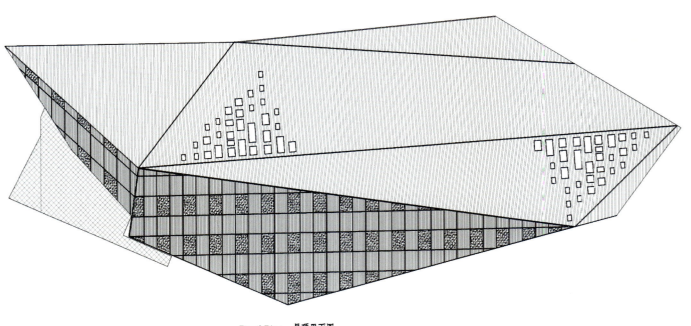

Roof Plan 屋顶平面图

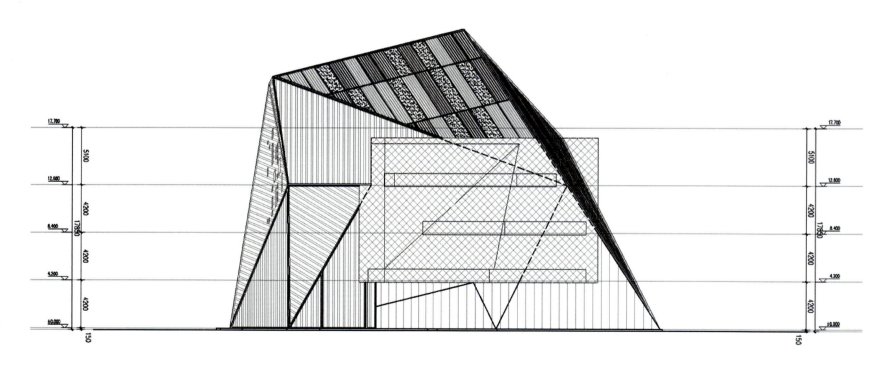

West Elevation　西立面

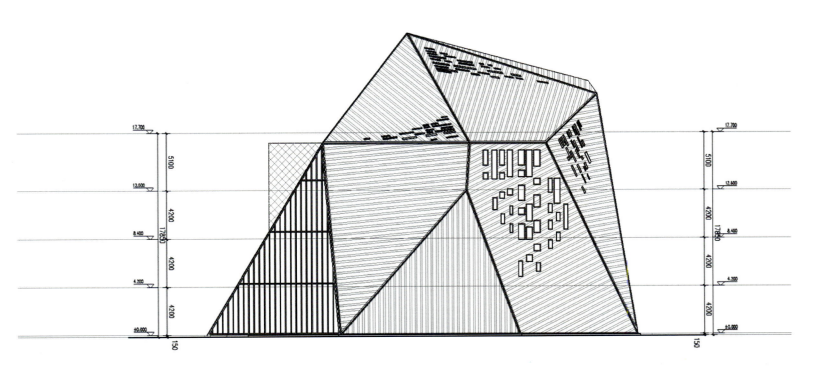

East Elevation　东立面

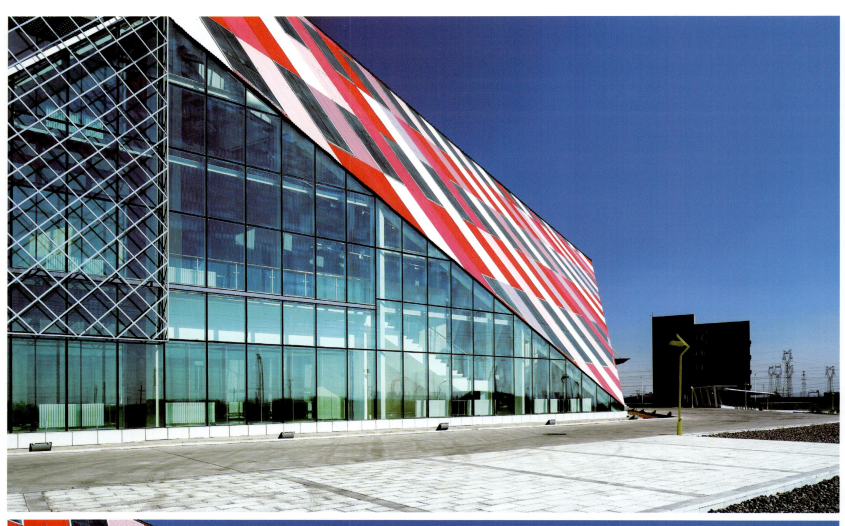

NORTHSTAR XINHE DELTA

北辰新河三角洲

ARCHITECT	LOCATION	AREA
THE JERDE PARTNERSHIP	Changsha, Hunan Province, China	520,000 m²

SUSTAINABLE & GREEN FEATURES 绿色特征

- *Use of friendly and low-energy materials* ／ 使用环保、低能耗材料

- *Green roof to reduce room temperature, recycle rainwater to irrigate the surrounding plants and lighten the heat island effect* ／ 绿色屋顶用于降低建筑室温、回收雨水用于浇灌周边绿地和减轻热岛效应

- *Water-saving equipments and energy-efficient heating and cooling systems* ／ 节水设备和节能加热制冷系统

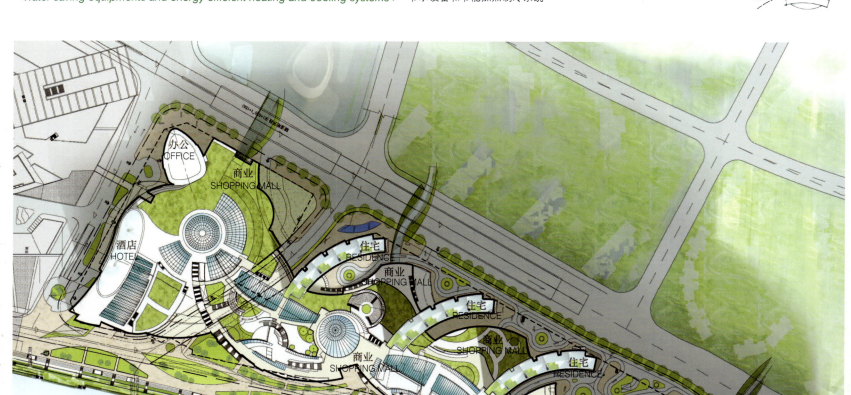

Rising out of the Xiang River, Northstar Xinhe Delta is a modern, waterfront lifestyle attraction for the city of Changsha. Symbolic of sails on the water's horizon, the sleek and contemporary design of the residential towers maximizes the westerly views to the river and Yuelu Mountain, while still capturing the ever precious southern sunlight. The project is designed to entice capture and generate social interaction within its pedestrian-oriented foundation. Whether it is a stroll along the "Boardwalk" lined with unique specialty shops and restaurants that take full advantage of the sweeping river views, or the interjected dazzling and entertaining array of activities, the waterfront will be enlivened with energy.

As sweeping towers of residences, hotel and offices soar above a rooftop park along the Xiang River, the clear interaction between built form and nature enhances the site's organic and fluid quality. Themed and inspired by the element of water, gardens, and city, Northstar Xinhe Delta's new and innovative concept of retail entertainment and mixed-use development is transformed away from the typical housing and retail environment, to deliver a dynamic destination to visit, shop, live, work, and play unlike any other. Complementing the adjacent cultural district, this new public arena will thrive with human activity and social interaction, establishing Northstar Xinhe Delta as the city's premier destination.

OFFICE TOWER
办公楼

HOTEL TOWER
酒店

RESIDENTIAL TOWER#1
住宅楼 1

RESIDENTIA
住宅

中庭
ATRIUM

影院
THEATER

RESIDENTIAL POOL DECK
住宅游泳区

RESIDENTIAL TOW
住宅楼 3

RESIDENTIAL TOWER#4
住宅楼 4

RESIDENTIAL TOWER#5
住宅楼 5

BRIDGE @ +39.0M

SOUTH ENTRY
南入口

RESIDENTIAL #5
DROP OFF
住宅楼 5 下客区

RESIDENTIAL #4
DROP OFF
住宅楼 4 下客区

RETAIL DROP OFF/ENTRY
购物区下客区 / 入口

RESIDENTIAL #3
DROP OFF
住宅楼 3 下客区

RESIDENTIAL TOWER#3
住宅楼 3

RESIDENTIAL TOWER#4
住宅楼 4

RESIDENTIAL TOWER#5
住宅楼 5

RESIDENTIAL POOL DECK
住宅游泳区

RESIDENTIAL POOL DECK
住宅游泳区

RESIDENTIAL POOL DECK
住宅游泳区

SOUTH ENTRY
南入口

OFFICE TOWER
办公楼

RESIDENTIAL TOWER#2
住宅楼 2

RESIDENTIAL TOWER#1
住宅楼 1

HOTEL TOWER
酒店

THEATER
影院

BRIDGE at +39.0M

BRIDGE at +39.0M

RESIDENTIAL DROP OFF
住宅楼 2 下客

RESIDENTIAL DROP OFF
住宅楼 1 下客

HORIZONTAL 4TH RD.
水平四路

RETAIL DROP OFF/ENTRY
购物区下客区 / 入口

东立面图

EAST ELEVATION

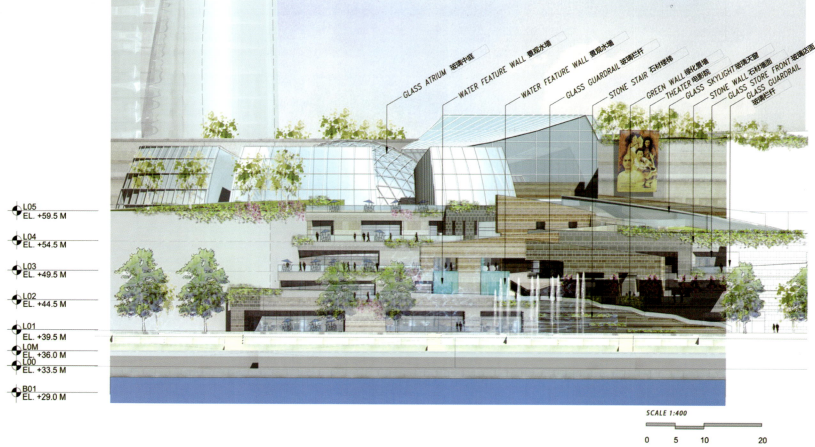

GLASS ATRIUM 玻璃中庭　WATER FEATURE WALL 景观水墙　WATER FEATURE WALL 景观水墙　GLASS GUARDRAIL 玻璃栏杆　STONE STAIR 石材楼梯　GREEN WALL 绿化景墙　THEATER 电影院　GLASS SKYLIGHT 玻璃天窗　STONE WALL 石材墙面　GLASS STORE FRONT 玻璃店面　GLASS GUARDRAIL 玻璃栏杆

L05
EL. +59.5 M

L04
EL. +54.5 M

L03
EL. +49.5 M

L02
EL. +44.5 M

L01
EL. +39.5 M

L0M
EL. +36.0 M

L00
EL. +33.5 M

B01
EL. +29.0 M

SCALE 1:400

0   5   10   20

1号住宅楼
RESIDENTAIL TOWER #1

STONE WALL
石材墙面

STAIR TO L02 楼梯

WATER FEATURE 水景

CONCRETE COLUMN (FACING T.B.D.)
混凝土柱

GLASS CURTAIN WALL 玻璃幕墙

STONE WALL 石材墙面

GLASS STORE FRONT 玻璃店面

RAMP TO L01 坡道

BRIDGES ON L01 AND L02
步行天桥

CINEMA
I-MAX

K T V

L05
EL. +59.5 M

L04
EL. +54.5 M

L03
EL. +49.5 M

L02
EL. +44.5 M

L01
EL. +39.5 M

L0M
EL. +36.0 M

L00
EL. +33.5 M

B01
EL. +29.0 M

SCALE 1:400

0   5   10   20

北辰新河三角洲项目地处长沙市，沿湘江拔地而起，极具现代气息，带给人们滨水生活的享受。高层住宅群的设计造型优美现代，从江面上看宛如船帆，最大限度地独揽西面河流与岳麓山景观，同时可坐享南面的宝贵光照。该项目的设计目的是激发人们的探索精神，在以行人为本的基础上，进行社交往来。人们可以漫步在"木板路"上，伴随着道路两旁独特的专卖店与餐厅，尽享整个江面的景色，也可以参与到各种意想不到、令人眼花缭乱的有趣的活动中。这些都使滨水区生机勃勃。

成片的住宅、酒店和写字间大楼沿着湘江林立，俯瞰着屋顶公园。建筑形式与大自然之间清晰的互动彰显出该地区的整体性与流动性。北辰新河三角洲以水、花园及城市为主题并从中获得灵感，它在零售娱乐与多用途发展方面观念新颖，已摆脱了传统的住宅及零售环境，打造出一个充满活力的场所，集游览、购物、居住、工作及娱乐于一体。这一新建的公共场所与邻近文化区相辅相成，它将随着人类活动及社交的不断发展而繁荣起来，将北辰新河三角洲打造成城市翘楚。

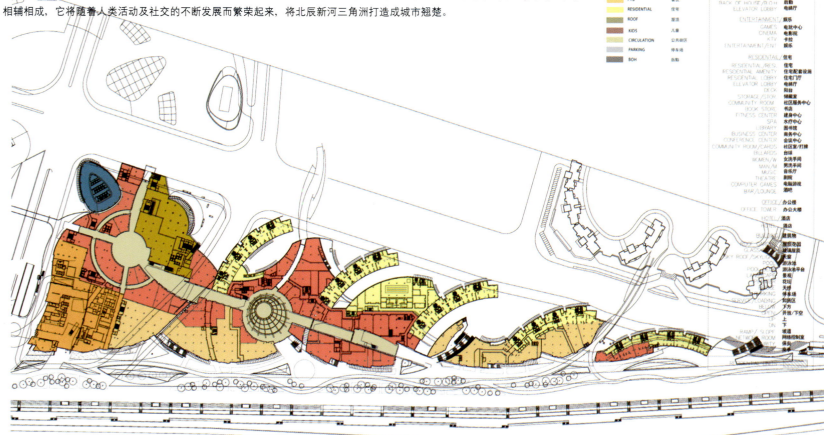

TEDA MSD H2
LOW CARBON BUILDING

泰达 MSD H2 低碳示范楼

CHINA GREEN BUILDING 3 STAR CERTIFIED, JAPAN GREEN BUILDING CASBEE CERTIFIED, UK BREEAM CERTIFIED, US LEED CERTIFIED

中国绿色建筑三星认证、日本 CASBEE 认证、英国 BREEAM 认证、美国 LEED 认证

SUSTAINABLE & GREEN FEATURES　绿色特征

- *Use of photovoltaic technology, solar energy, renewable and recycle materials* ／ 采用光伏技术、太阳能、可再生和可回收材料
- *Highly efficient and energy-saving rainwater and grey water recycling system, lighting system, water supply and draining system and ventilation system* ／ 高效节能的雨水及中水回收系统、照明系统、供水与排水系统和通风系统

ARCHITECT	**CLIENT**	**LOCATION**	**AREA**
Atkins	Tianjin TEDA	Tianjin, China	21,158 m²

Located in Tianjin TEDA Modern Service Industrial District (MSD), H2 Low Carbon Building is conceived as a demonstration project and research platform for green building technologies.

Nickamed by the design team "ecological sandwich", the project has slim rectilinear appearance and performance-inspired façades. Ecological sandwich means that Core cube is placed on both the north and south sides. The outermost layer of the south façade is the solar PV glass which is independent of the building's main structure. The third floor and the south façade above the third floor are designed as double layer breathing wall, and the north façade is designed as the double hollow glass wall to meet the "low carbon and energy-saving" need.

H2 low carbon building is likely to become the first low carbon building attaining four green building certifications at the same time, including China 3 Stars, Japan CASBEE, UK BREEAM and the US LEED. The whole building utilizes various green technologies from the basic to the most advanced, such as the recycling of the rain, photovoltaic technology, solar energy, etc. These technologies make low carbon and energy saving possible and create a comfortable inner environment.

The low carbon technology used in the building includes five aspects: architecture and structure, renewable energy, electrical and lighting design, plumbing and sanitary engineering, and HVAC engineering. As a low carbon showcase, apart from using new materials and new technologies reasonably and exploringly, the project is an attempt to integrate low carbon design with a smart, attractive architectural expression.

五大体系 **24项技术**

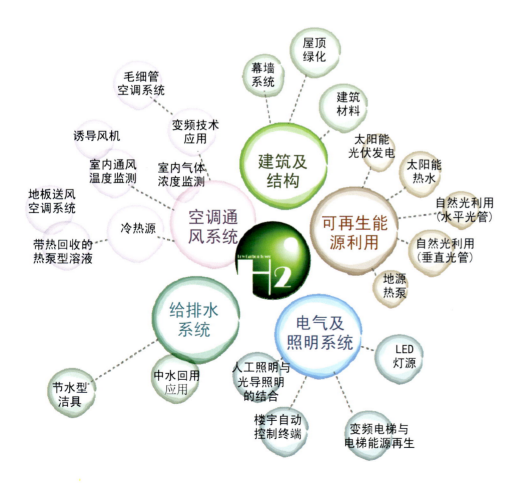

屋顶
绿化

幕墙
系统

建筑
材料

毛细管
空调系统

变频技术
应用

诱导风机

室内气体
浓度监测

室内通风
温度监测

地板送风
空调系统

冷热源

带热回收的
热泵型溶液

太阳能
光伏发电

太阳能
热水

自然光利用
（水平光管）

自然光利用
（垂直光管）

地源
热泵

**建筑及
结构**

**空调通
风系统**

**可再生能
源利用**

**给排水
系统**

**电气及
照明系统**

LED
灯源

节水型
洁具

中水回用
应用

人工照明与
光导照明
的结合

楼宇自动
控制终端

变频电梯与
电梯能源再生

•建筑及结构
•可再生能源利用
•电气及照明系统
•给排水系统
•空调通风系统

总节能耗 **30%**

融入建筑的低碳技术

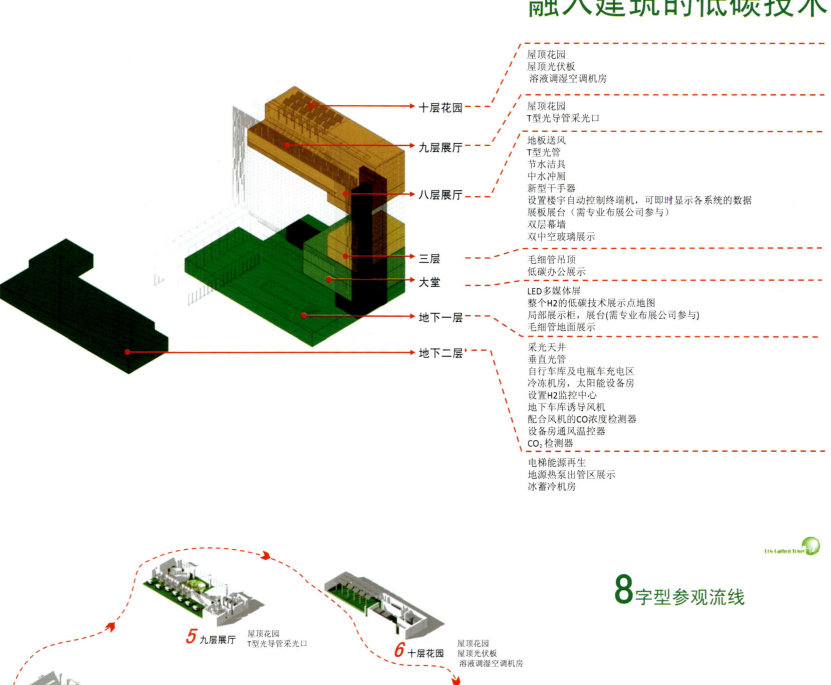

十层花园 —— 屋顶花园
屋顶光伏板
溶液调湿空调机房

九层展厅 —— 屋顶花园
T型光导管采光口

八层展厅 —— 地板送风
T型光管
节水洁具
中水冲厕
新型干手器
设置楼宇自动控制终端机，可即时显示各系统的数据
展板展台（需专业布展公司参与）
双层幕墙
双中空玻璃展示

三层 —— 毛细管吊顶
低碳办公展示

大堂 —— LED多媒体屏
整个H2的低碳技术展示点地图
局部展示柜，展台(需专业布展公司参与)
毛细管地面展示

地下一层 —— 采光天井
垂直光管
自行车库及电瓶车充电区
冷冻机房，太阳能设备房
设置H2监控中心
地下车库诱导风机
配合风机的CO浓度检测器
设备房通风温控器
CO_2检测器

地下二层 —— 电梯能源再生
地源热泵出管区展示
冰蓄冷机房

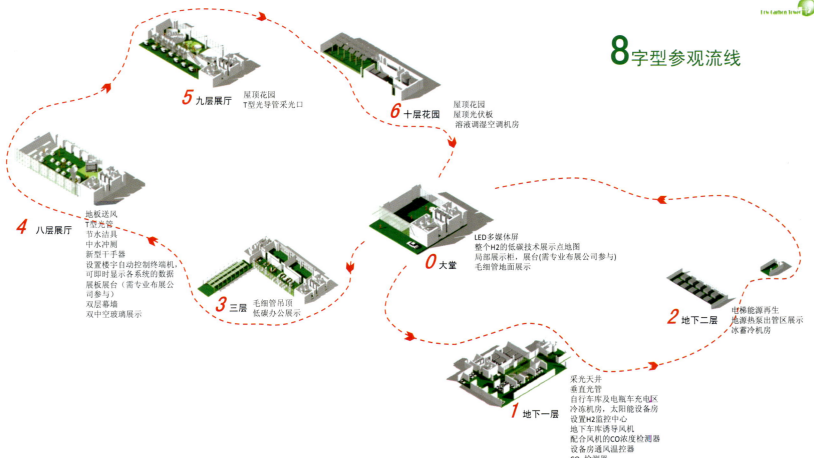

8字型参观流线

5 九层展厅　屋顶花园
T型光导管采光口

6 十层花园　屋顶花园
屋顶光伏板
溶液调湿空调机房

4 八层展厅　地板送风
T型光管
节水洁具
中水冲厕
新型干手器
设置楼宇自动控制终端机，可即时显示各系统的数据
展板展台（需专业布展公司参与）
双层幕墙
双中空玻璃展示

3 三层　毛细管吊顶
低碳办公展示

0 大堂　LED多媒体屏
整个H2的低碳技术展示点地图
局部展示柜，展台(需专业布展公司参与)
毛细管地面展示

2 地下二层　电梯能源再生
地源热泵出管区展示
冰蓄冷机房

1 地下一层　采光天井
垂直光管
自行车库及电瓶车充电区
冷冻机房，太阳能设备房
设置H2监控中心
地下车库诱导风机
配合风机的CO浓度检测器
设备房通风温控器
CO_2检测器

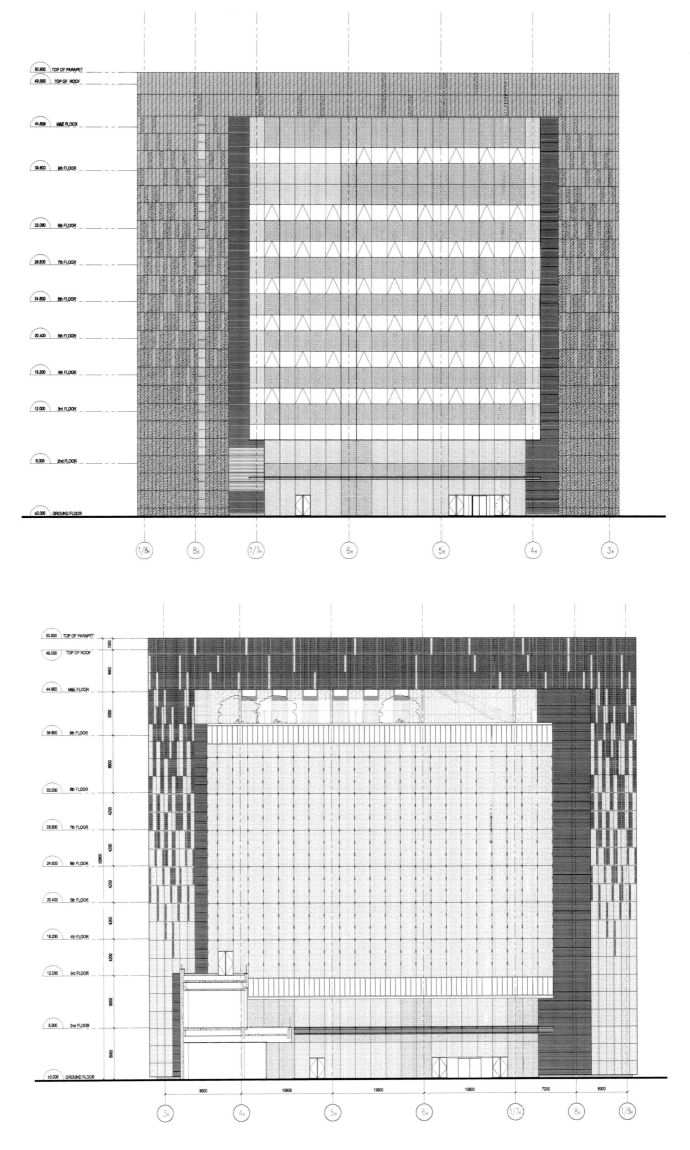

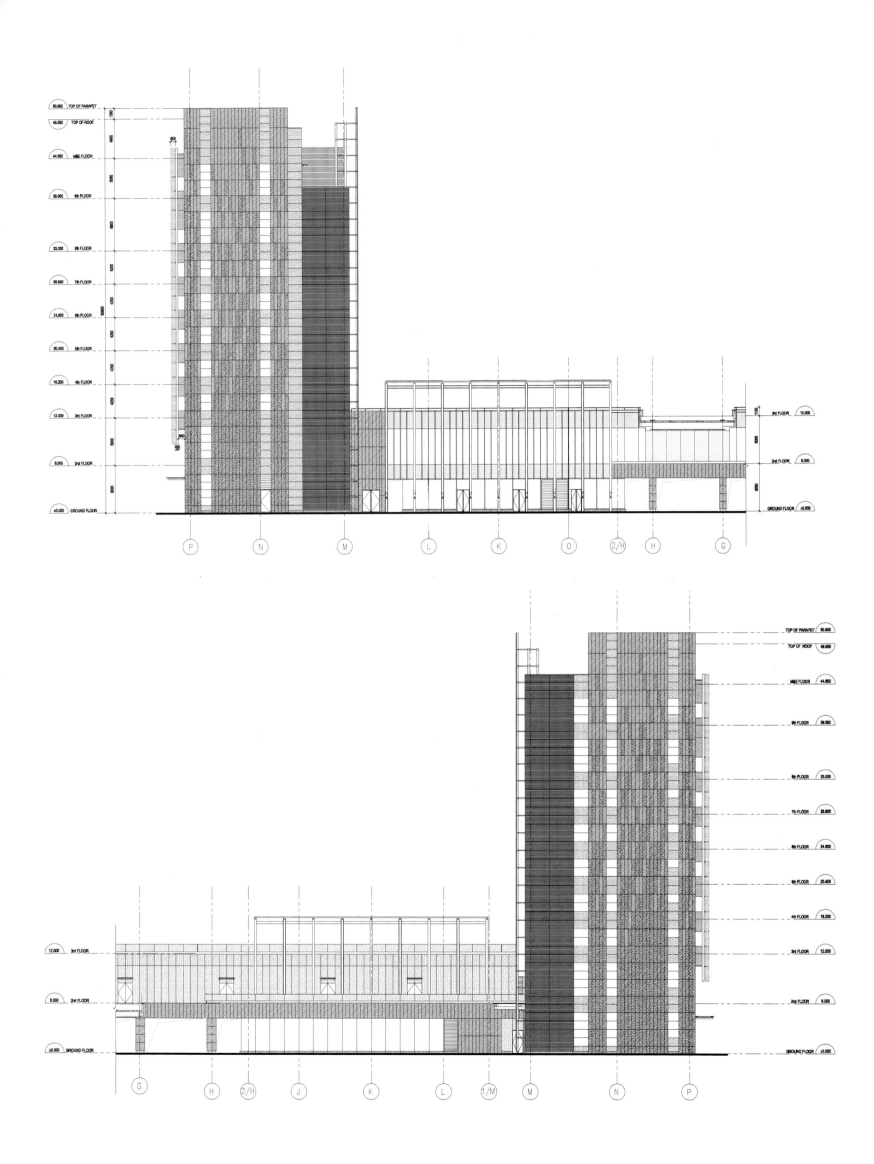

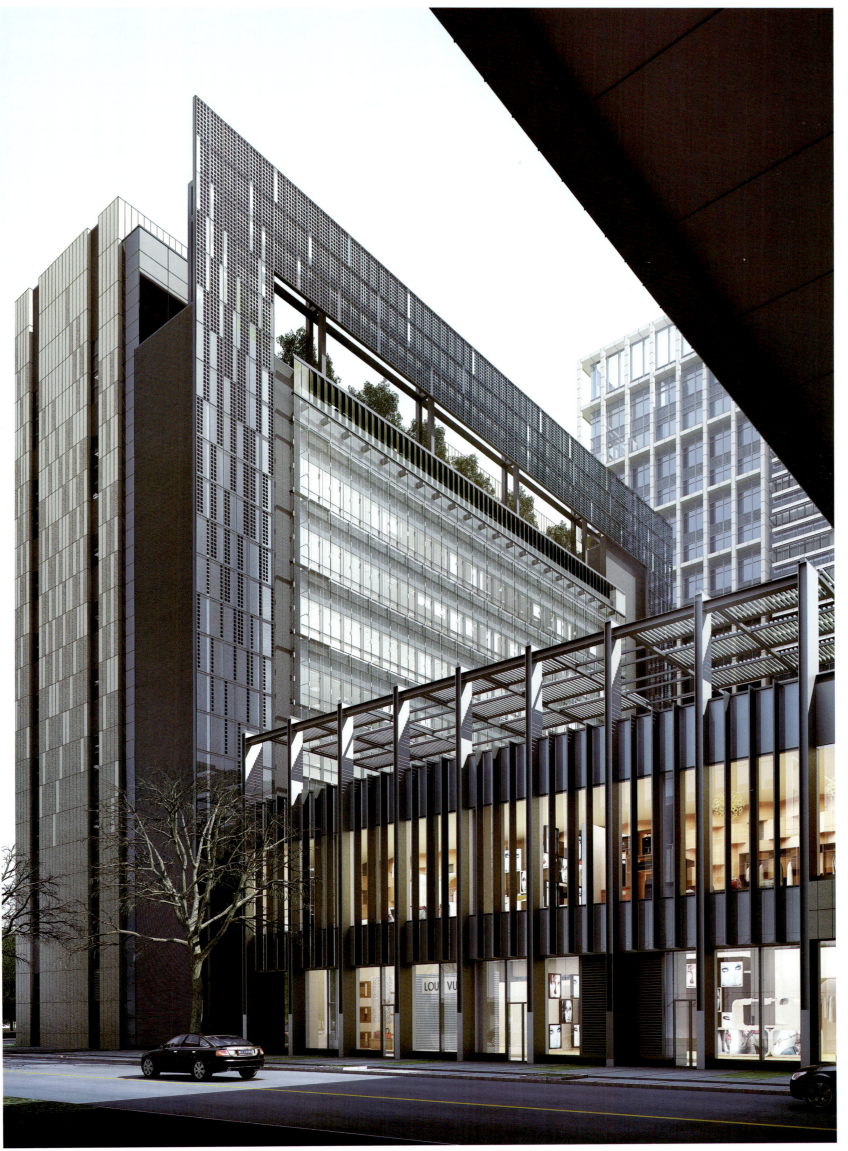

H2 低碳示范楼位于天津泰达现代服务产业园区内，具有示范功能，为高新低碳技术提供展示平台。

基于该建筑物的瘦长体量和独特的高性能幕墙体系，设计团队将其称为"生态三明治"，即将核心筒设置在南、北两侧。南立面最外层为光伏玻璃，独立于建筑主体结构之外。三层以上南立面设计为双层呼吸式幕墙，北立面为双中空玻璃幕墙，充分考虑了低碳节能的要求。

H2 低碳示范楼将有机会成为世界上首个同时通过四项绿色建筑认证的低碳建筑，分别为中国三星级认证、日本 CASBEE 认证、英国 BREEAM 认证、美国 LEED 认证。整个建筑采用了从最基础的雨水回收利用到最先进的光伏技术，太阳能利用等多种绿色技术，既做到了低碳节能，又创造出舒适的内部环境。

建筑中采用的低碳技术，主要包括五个体系：建筑及结构、可再生能源利用、电气及照明系统、给排水系统、空调通风系统。作为低碳示范楼，除了理性地、探索地运用新材料、新技术，整个建筑还将低碳设计与现代、醒目的建筑形态有效地整合在一起。

QIAOXIANG VILLAGE, SHENZHEN

深圳市
侨香村

ARCHITECT
Capol

LOCATION
Shenzhen, Guangdong Province, China

AREA
127,449.85 m² (Gross Site), 500,080 m² (Gross Built)

SUSTAINABLE & GREEN FEATURES 绿色特征

- *Use of local and recyclable materials* ／ 使用当地的可回收材料
- *Use of solar lighting system, solar hot water system, rainwater collecting and recovery system, and grey water recycling system* ／ 使用太阳能照明系统、太阳能热水系统、雨水回收利用系统和中水处理系统

The project is situated in the eastern part of Antuo Hill of Futian District, Shenzhen, with a total land area of 127,449.85 square meters and a total gross floor area of 500,080 square meters. Based on the concept of urban design, it makes maximal use of the environmental resources of this land, building a simple, comfortable and high-class living space.

The façade, with simple shape, combination of grey and white in color, appropriate details, harmonious coexistence with the surroundings, best embodies the basic characteristics of affordable housing, but all at the same time shows a sense of dignity. The overall design seems clear and simple. The residential tower is the landmark of this area with dark bluish grey metope in contrast with beige framework. The thick and thin lines in grey and white add innovation to the sedate design. Its modern style

can best represent the spirit of Shenzhen. The roof and floor of basement have beamless floor system, which can decrease the floor height of the basement in order to reduce earth excavation, enhance the speed of construction and lower the construction cost.

High strength concrete is used in the frame column and frame-supporting column of this project so as to reduce the size of the cross section of the column and increase the area utilization. High strength steel bars are used as much as possible to minimize the total use of steel bars of any kind. HRB335 is seen in concrete column, HRB400 in the wall beam, and cold-rolled deformed bars or cold-rolled ribbed bars are used in the roof and floor of basement and civil air defense roof. Aerated concrete blocks are seen in the exterior wall and light wall boards or aerated concrete block are used in the interior wall to reduce structural weight and decrease construction cost.

Water Supply and Drainage Design

Eco-friendly Materials: All the equipments selected are recommended products by the National Ministry of Construction. Facilities to abolish poisonous gas or sound pollution are used when such pollutions occur. Domestic sewage is treated through the septic system before it is discharged.

Energy-saving: All the equipments are energy-saving, including the converter speed control water pump.

Water-saving: Such sanitary ware and water supply fittings as toilet of two tranches, time-lapse self-closing flush, urinals, toilets are all water-saving. Rain water is collected for irrigation and road sprinklings. Tap water can only supply for domestic use. Compared with other water-saving ways , the rainwater utilization project has three advantages: "reuse, pollution reduction and flood elimination". Structures designed for regulation and storage, purification and permeation mainly include storage pool of the basement, greenbelt, sewer, leaky pipe and soil filter.

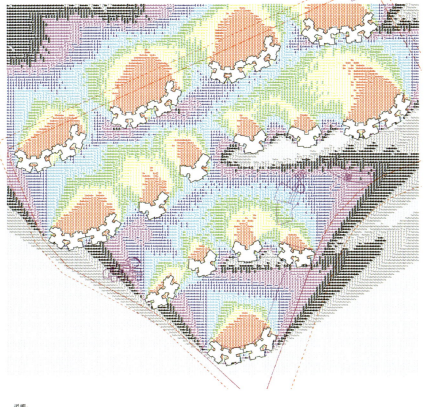

说明：
1、本计算采用软件为《天正 VI》
2、图中所示数字为深圳地区，大
寒日，在建筑的综合影响下，
底层窗台的日照时间。
3、有效时间为8: 00～16: 00

Sunlight Analysis
日照分析图

本项目位于深圳市福田区安托山东片区，总用地面积127449.85平方米，总建筑面积500080平方米。本方案从城市设计的观点出发，最大化地挖掘该地块的环境资源，营造简洁、舒适、高品位的生活空间。

立面设计以简洁的造型、灰与白的协奏、恰当的细部，与周围环境和谐共存，既体现经济适用房的基本特征，又不失庄重。住宅塔楼以深蓝灰色主墙面与米黄色的构架形成本小区的特征印象符号，整体设计简洁明了。墙身中穿入灰色、白色的粗线条和细线条，在沉稳中体现创新。以现代风格为导向，体现深圳精神。地下室顶板、底板采用无梁楼盖体系，降低地下室层高以减少土方开挖，提高施工速度，降低造价。

本工程框架柱及框支柱混凝土采用高强度混凝土，以减小柱截面尺寸，达到提高面积利用率的目的。墙柱钢筋采用HRB335，墙梁钢筋采用HRB400，地下室底板、顶板、人防顶板等采用冷轧变形钢筋或冷轧带肋钢筋，尽可能在受力较大部位使用高强度钢筋以减小含筋量，节省钢筋。外墙采用蒸养加气混凝土砌块，内墙采用轻质墙板或蒸养加气混凝土砌块，以减轻结构重量，降低造价。

给水排水设计

环保材料：所有设备、器材均选用建设部推荐产品，凡对环境产生废气、噪音的设备，均设置相应的除废气、消声等处理设施，生活污水经化粪池处理后再外排。

节能：所有设备均选用节能型，如采用变频调速给水泵。

节水：卫生器具及给水配件均采用节水型，如采用节水型两档水箱大便器，延时自闭冲洗龙头，延时自闭冲洗小便器，延时自闭冲洗大便器等。收集雨水用于绿化喷灌、浇洒道路，自来水只供应给居民生活用水，雨水利用工程的"回用、减污、消洪"三重效果与其他节水方式相比，具有一定的优势。在调蓄、净化和渗透方面选取的构筑物主要包括：地下室储存水池、绿地、渗沟、渗管、土壤滤池等。

法定图则拟调整为9年制学校用地

北　环　路

安　托　山　九　路

金地香蜜山

莲花四路

安　托　山　七　路　香　路

高压走廊保护范围

● 受北环路噪音影响的住宅单元
✳ 噪声监测点

防噪措施一：

规划及建筑设计上的防噪措施：

1. 通过小区规划使住宅尽量远离交通干道，其中临北环路住宅离北环路73米至86米。
2. 通过建筑单体设计，尽量减少受北环路噪声影响的住宅单元，且每单元仅卫生间、厨房及次卧受到影响。
3. 加强主干道与住宅间隔离带的绿化，从传播途径降低噪声污染，将噪声污染降低到最低水平。

防噪声设计（一）——规划与建筑防噪

北　环　路

安　托　山　九　路

安　托　山　八　路

莲花四路

安　托　山　七　路　面

● 组团绿地
● 入口广场
● 缓坡景观绿化带
● 周边景观绿化带
● 中心服务景观绿化带

环境景观绿化分析

1. 高绿地率，小区绿化占用地面积的六成左右，充分体现了"绿色小区"的主题和立意。
2. "点、线、面"的有机结合，点：各种植物的种栽；线：周边绿化带；面：公共绿地、组团绿地、生态停车场的设置。
3. 多层次的景观轴串连，通过广场的相接，给人移步换景的视觉享受。

环境景观绿化分析图

MUSEUM OF CONTEMPORARY ART & PLANNING EXHIBITION, SHENZHEN, CHINA

中国深圳当代艺术馆和规划展览馆

SUSTAINABLE & GREEN FEATURES　绿色特征

- *Use of energy saving façade materials* ／　表皮使用节能材料
- *Large use of solar energy and geothermal energy* ／　大量使用太阳能和地热能
- *Highly efficient ventilation system and combined heat and power generation system* ／　高效节能的通风系统和热电联产系统

ARCHITECT
COOP HIMMELB(L)AU, Wolf D. Prix / W.
Dreibholz & Partner ZT GmbH

PROJECT ARCHITECT
Angus Schoenberger

DIGITAL PROJECT TEAM
Angus Schoenberger, Matt Kirkham, Jasmin
Dieterle, Jonathan Asher, Jan Brosch

DESIGN PRINCIPAL
Wolf D. Prix

DESIGN ARCHITECT
Quirin Krumbholz, Jörg Hugo, Mona BayrProject

LOCATION
Shenzhen, Guangdong Province, China

PROJECT PARTNER
Markus Prossnigg

COORDINATION
Veronika Janovska

AREA
21,688 m² (Site), 110,000 m² (Gross Floor)

The Museum of Contemporary Art & Planning Exhibition (MOCAPE), part of the master plan for Shenzhen's new urban center, the Futian Cultural District, establishes itself as a new attraction in Shenzhen's fast growing urban fabric. The project is conceived as the synergetic combination of two institutions, the Museum of Contemporary Art (MOCA) and the Planning Exhibition (PE), whose various programmatic elements, although each articulated according to their functional and performative requirements, are merged in a monolithic body enveloped by a multifunctional façade.

The transparency of this façade and the interior lighting concept allow for a view from outside through the exterior envelope deep into the volume of the space, thereby particularly accentuating the shared entrance and circulation space between the two museum volumes. At the same time, the

building skin also allows unhindered view from the inside on the cityscape while giving the visitor the impression of being in a pleasantly shaded outdoor area – an impression enhanced by very wide spans which allow for completely open, column-free and flexible exhibition halls with heights ranging from 6 to 17 meters.

Upon entering the large interstitial space between the two museum volumes, the visitors can, via ramps and escalators, reach the main level, where a kind of public plaza serves as the circulation hub for the whole complex and as orientation and starting point for museum tours. From here also many auxiliary public and private facilities are accessible, including cultural services, a multi-functional hall, several auditoria, a library, a cafe, a book store, and a museum store.

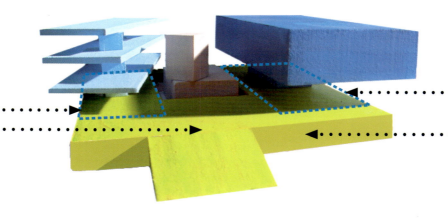

Planning Exhibition Platforms
规划展览平台

Contemporary Art
Exhibition Box
当代艺术展览盒

Additional Exhibition Space
for Planning Exhibition
规划展览馆的附加展览区
Open Plaza +10m
开放式广场（+10m）

Additional Exhibition Space
for Contemporary Art
现代艺术附加展览区

Combined service base for both museums
 Administration
 Shops
 Multifunctional Space
两个博物馆的联合服务基地
行政
商店
多功能区

■ Contemporary Art Exhibition Box 当代艺术展览盒
■ Planning Exhibition Platforms 规划展览平台
■ Open plaza 开放式广场

Program 图解

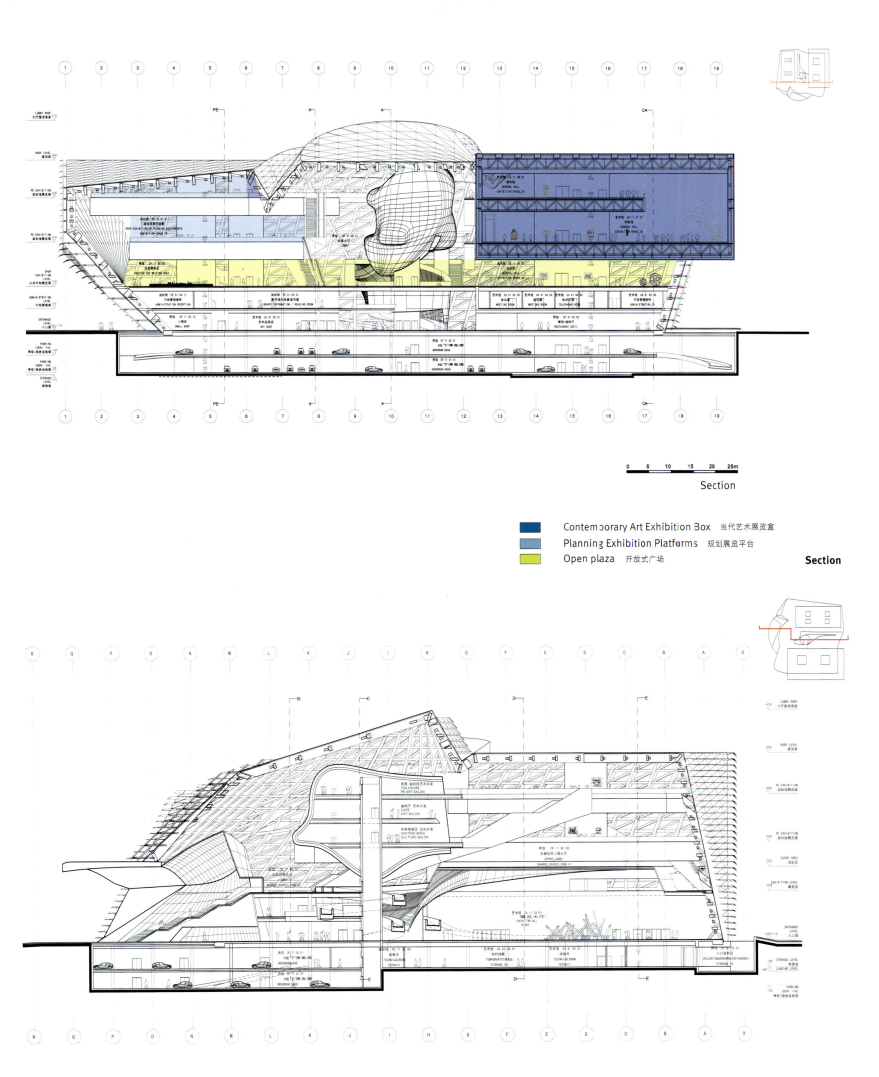

Section

- ◼ Contemporary Art Exhibition Box　当代艺术展览盒
- ◼ Planning Exhibition Platforms　规划展览平台
- ◼ Open plaza　开放式广场

Section

0　5　10　15　20　25m

5　10　15　20　25m

比例 **SCALE | 1:500**

A-A 剖面 **| SECTION A-A**

设计图 **| PLANS**

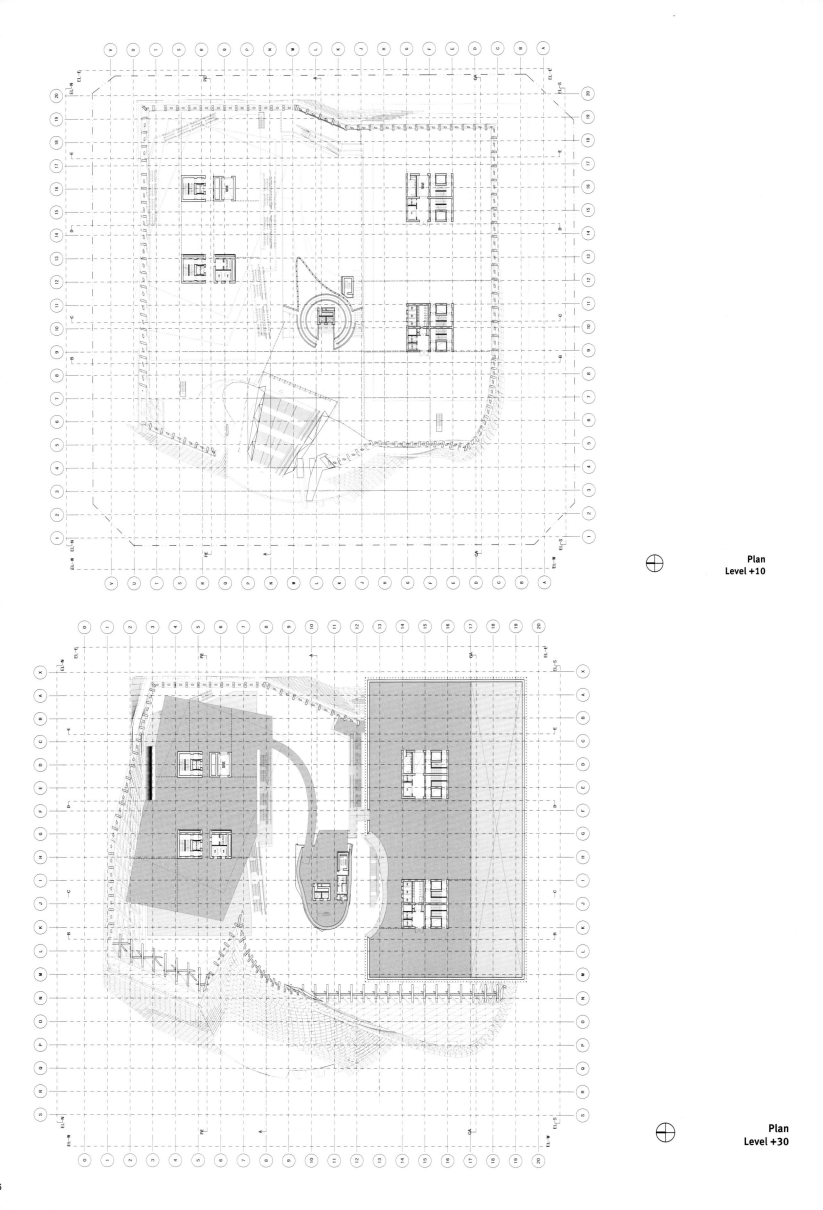

Plan
Level +10

Plan
Level +30

当代艺术馆和规划展览馆是深圳新市中心——福田文化区总体规划的一部分，也是深圳快速发展的城市结构中全新的魅力点。该项目的构想是将当代艺术馆和规划展览馆结合起来。这些各具风格的元素尽管有着不同的功能和表达诉求，却在多功能外墙的包裹下形成了一个庞大统一的整体。

透明的外墙和内部的照明设计使人们可以从外部底层空间看到大楼的内部，因此设计格外强调两个场馆的共享入口和流通空间。同时，访客在大楼内部也可以透过建筑幕墙一览无余地观赏到外面的城市风光，给人一种置身于遮阴舒适的室外空间的感觉。场馆没有围栏的限制，只有 6 米到 17 米不等的灵活的展览墙，赋予场馆完全开放的宽阔视野，使访客更加舒适自在。

访客可以选择走坡道或乘坐自动扶梯，通过两个场馆之间的空地来到主场馆。那里的公共广场是整个综合设施的枢纽，也是场馆之行必经的起点。从这里出发可以看到许多公共附属设施和专用设施，其中包括文化服务区、多功能大厅、几个会议厅、图书馆、咖啡厅、书店和展览品商店。

INDEX

索引

RONALD LU & PARTNERS

Ronald Lu and Partners is an award-winning architecture and interior design practice dedicated to the delivery of world-class projects and green built environments across the globe. Founded over 35 years ago, we have consistently pioneered sustainable architecture and are recognised as industry leaders. We have over 500 staff across our Hong Kong headquarters and four mainland China offices, and more than 70% of our accredited professionals have achieved BEAM Pro or equivalent international standard, which drives sustainability throughout our firm. We embrace aholisticapproach and are committed to excellence in design and construction, as well as maximising opportunities for greater social, economic and environmental benefits for communities. Our expertise in sustainability is integrated with our broad spectrum of design services including master planning, new build, interior design, urban regeneration and architectural research. We are delivering projects ranging from BEAM Plus assessmentstandard (and national/international equivalents), up to the state-of-the-art Zero Carbon building.

COOP HIMMELB(L)AU

COOPHIMMELB(L)AU
Wolf D. Prix & Partner ZT GmbH

COOP HIMMELB(L)AU was founded in Vienna in 1968 and has since been operating under the direction of Wolf D. Prix, CEO and design principal in the fields of art, architecture, urban planning, and design. Another branch of the firm was opened in the United States in 1988 in Los Angeles.

The company's most well-known international projects include the Falkestrasse attic conversion in Vienna, the multifunctional UFA Cinema Center in Dresden, the BMW Welt in Munich, as well as the Dalian International Conference Center in China.

Projects currently under construction include the Musée des Confluences in Lyon, France and the European Central Bank (ECB) in Frankfurt am Main.

COOP HIMMELB(L)AU currently employs over 150 people from 19 different countries.

KOKAISTUDIOS

Kokaistudios was founded in Venice in 2000 by Italian architects Filippo Gabbiani and Andrea Destefanis. Established with the dream to create a collaborative office of talented architects associated together to work on challenging and interesting projects on a world-wide basis the firm started with offices in Venice, Copenhagen and Hong Kong,China. From these early expansions the firm developed a multi-cultural and multi-disciplinary approach to projects which helped lead to the establishment in 2002 of the Shanghai office which today is the heart of Kokaistudios operations worldwide. Group of over 30 architects and interior designers are united in developing interesting and innovative commercial, institutional and hospitality projects in the field of architecture, heritage architecture, and interior design.

HELLER MANUS ARCHITECTS

Founded over 25 years ago, Heller Manus Architects is an internationally recognized, San Francisco-based architecture and urban design firm. Jeffrey Heller FAIA, President of the firm and council member of Copenhagen-based international Green Growth Leaders, and Clark Manus, FAIA, CEO of the firm and the 87th President of the American Institute of Architects, have dedicated more than twenty-five years to developing a diversified, client-oriented firm that is a design and sustainability leader in the profession. Their depth of experience encompasses a wide range of major new construction and historical renovation projects including mixed-use, commercial, residential, hospitality, civic, rehabilitation and adaptive re-use, performance facilities, sustainability, transportation, master planning and urban design.

RTKL ASSOCIATES INC.

A worldwide architecture, engineering, planning and creative services organization. Part of the ARCADIS global network since 2007, RTKL Associates Inc. specializes in providing its multi-disciplinary services across the full development cycle to create places of distinction and designs of lasting value. RTKL Associates Inc. works with commercial, workplace, public and healthcare clients on projects around the globe.

BLVD INTERNATIONAL INC.

BLVD international inc. was founded in Ontario, Canada in October 2001. The design company was established by several Canadian Architects and Interior Designers who shared a common interest in working in the Chinese market. After ten years of professional practice in China, BLVD now specializes in the field of Architectural Design, Urban Planning, Landscape Design and Interior Design including Resort Planning, high-end Hospitality and Office developments, Model Suites and Club Houses.

JOHNSON PILTON WALKER

JPW
JOHNSON PILTON WALKER

JOHNSON PILTON WALKER is a Sydney based design studio with major built works in architecture, planning, urban design, landscape architecture, interior design and exhibitions both in Australia and internationally.

With more than 50 design professionals in the studio, their multidisciplinary approach enables them to innovate from project conception, through design development, detailing, documentation to completion and ongoing asset management.

Over the past 10 years their projects have been awarded many prizes and awards, including major international awards in China, New Zealand and the USA.

They are driven by strong environmental principles – to create places where people are happy to work, live and play; that are efficient to run; have low energy usage; good urban connections and excellent design.

MING LAI

Ming Lai obtained his master degree from Illinois Institute of technology in 1997. He is one of the 3 recipients for the Architect Club of Chicago Award in 1996. Ming Lai Architects Inc. (MLAI) was established in 2008. This fledging architectural company is headed by Ming C. Lai, who brings with him a distinguished international portfolio. Prior to founding Ming Lai Architects Inc., he spent 12 years with GP, an established American architecture company as senior architect and project leader, playing a pivotal role as the designer and project manager for many large international projects. In China, Ming Lai has led and involved in more than ten projects in Shanghai, Beijing, Suzhou, Chengdu, Nanjing, Tianjin and many other cities. His profound understanding of the design and needs of offices, hotels and mixed-use development is based on his years of a combination of theory and practice. Thus, he wins the unqualified approval of his clients with regards to his professional capacities and rich experiences.

PAUL NORITAKA TANGE

Paul Noritaka Tange began his architectural career upon receiving his Master in Architecture from Harvard University, Graduate School of Design in 1985. That same year he joined Kenzo Tange Associates, the architectural firm headed by his father, well known international architect, Kenzo Tange. Paul became President of Kenzo Tange Associates in 1997 and founded Tange Associates in 2003. Tange Associates, headquartered in Tokyo, Japan, has worked worldwide and offers a full range of architectural and urban design and planning services. At this time, Tange Associates has close to 40 on-going projects in ten countries. The firm's extensive international experience enables it to work effectively in all cultures. Its long standing associations with local architectural firms and its familiarity with local building practices is invaluable to Tange Associates' ability to efficiently undertake small to large scale projects in urban as well as rural areas in all parts of the world. Paul himself exemplifies the international element of his practice. Born in Tokyo, Japan, and educated in Japan, Switzerland and the US, he is a registered architect of both Japan and Singapore.

ARQUITECTONICA

ARQUITECTONICA

Arquitectonica is a full-service architecture, interior design and planning firm that began in Miami in 1977 as an experimental studio. Led by Bernardo Fort-Brescia, FAIA, and Laurinda Spear, FAIA, ASLA, the studio has evolved into a worldwide practice, combining the creative spirit of the principals with the efficiency of delivery and reliability of a major architectural firm. Its affiliated firm, ArquitectonicaGEO, provides landscape architecture services.

Today Arquitectonica has a practice across the United States directed from regional offices in Miami, New York and Los Angeles. Arquitectonica's international practice is supported by their European regional office in Paris; Asian regional offices in Hong Kong, China, Shanghai, China, Manila; Latin American regional offices in Lima and São Paulo; and the Middle East office in Dubai.

9TOWN-STUDIO

9town-studio was set up in January, 2002, with 2 branches in Shanghai and Suzhou. It has the Class-A design certification awarded by Housing and Urban-Rural Development and its business includes urban design, architecture design, landscape design and interior design. Over 10 years, 9town-studio always advocates the design concepts and professional aspirations with combination of art, technology and life, in order to make design open and become a real game which feels space, design life and construct future.

MAYU ARCHITECTS+

MAYU architects+, founded in 1999 as Malone Chang Architects, later as Malone Chang and Yu-lin Chen Architects, is an interdisciplinary practice based in Kaohsiung, Taiwan, China. The scope of MAYU architects+'s works spans from large scale civic projects to small scale ones. The studio recognizes the importance of teamwork in the creation of architecture, while maintaining the individuality and artistry of each project. Therefore an architectural approach balancing the interdisciplinary cooperation and artistic inspiration is applied. The studio's conception to a project avoids presumption of fixed style and dogmatism, in favor of a dynamic and organic process. The dialogue between site, context, programs, time, materials, users, and clients constantly contributes to the forming of architecture. Organizational logic of physical buildings, holistic experience of spaces, and the changing characteristics of materials are priorities of the studio.

C85 M21 Y0 K0

Atkins is one of the world's leading design consultancies. They have the breadth and depth of expertise to respond to the most technically challenging and time-critical projects and to facilitate the urgent transition to a low carbon economy. Their vision is to be the world's best design consultant.

Whether it's the architectural concept for a new supertall tower, the upgrade of a rail network, master planning a new city or the improvement of a management process, they plan, design and enable solutions.

With 75 years of history, 17,700 employees and over 200 offices worldwide, Atkins is the world's 13th largest global design firm (ENR 2011), the largest global architecture firm, the largest multidisciplinary consultancy in Europe and UK's largest engineering consultancy for the last 14 years. Atkins is listed on the London Stock Exchange and is a constituent of the FTSE 250 Index.

HYHW ARCHITECTURE CONSULTING LTD

HYHW

HYHW Architecture Consulting LTD is registered in London and Beijing. The company is multidisciplinary, specializing in architecture, planning, urban design, and interior design. Projects include commercial buildings, shopping malls, urban planning, urban design, public buildings, tourism, sports and cultural recreation, education, and residential developments. We value the demand of the clients and final users, and guarantee high-quality design.

CAPOL

CAPOL 華陽國際

Capol was established in November 2000 with headquarters in Shenzhen and branch offices in Guangzhou, Shanghai, Changsha, Chongqing, and Hong Kong, China. With its Class-A Comprehensive Architectural Design Qualification and Class-B Municipal Planning Design Qualification, Capol now boasts more than 1,500 employees. Honored as one of the Top 10 Privately Owned Design Enterprises in China, as an Annual Industry Leader, as the Most Competent Company, and as an Outstanding Private Company in Survey and Architectural Design, Capol enjoys a stellar reputation among well-known real estate developers, such as Vanke, CRC, Huawei, Longfor, Poly, China Merchants, CITIC, AVIC, Shum Yip, Kingkey, Gemdale, Excellence, Franshion, HWL, Mapletree and so on, at the same time Capol engaged in fruitful cooperation with SOMFoster, RTKL, Nikken Sekkei, CALLISON, BENOY, TFP, ARUP, JERDE and other top design companies in the world.

Founded in San Diego, California in 1977, JWDA is dedicated to excellence in urban design, planning, architecture, landscape and interior design. Its expertise spans a full spectrum of professional design services in hospitality, commercial, residential, educational and recreational communities, as well as data centers and urban complexes. JWDA committed to research in all new advanced building technologies and continuous strives to create a better environment through sustainable designs and innovative solutions. JWDA is a member of AIA (the American Institute of Architects), NCARB (National Council of Architectural Registration Boards) and ULI (Urban Land Institute). It is chaired by an acclaimed architect Joseph Wong, FAIA. He earned his Bachelor of Arts with Honors and Master of Architecture from the University of California, Berkeley; and Master of Landscape Architecture from Harvard Graduate School of Design. He served as a Board of Director for the Center City Development Corporation (CCDC) for the City of San Diego from 1991 to 1994; and as well as a Regional Chair and Board Member of the Harvard Graduate School of Design Alumni Council.

THE JERDE PARTNERSHIP

The Jerde Partnership is a visionary architecture and urban design firm that creates distinct places for people that deliver memorable experiences. Over 1 billion people visit Jerde Places annually. Founded in 1977, the firm has pioneered "placemaking" throughout the world with projects that provide lasting social, cultural and economic value, and promote further investment and revitalization. Based in a design studio in Los Angeles with project offices in Hong Kong, Shanghai and Seoul, Jerde takes a signature, co-creative approach to design and collaborates with private developers, city officials, specialty designers and local executive architects to realize the vision of each project. The firm has received critical acclaim from the Urban Land Institute, International Council of Shopping Centers, and American Institute of Architects. To date, over 110 Jerde Places have opened in diverse cities, including Atlanta, Budapest, Hong Kong(China), Istanbul, Las Vegas, Los Angeles, Osaka, Rotterdam, Seoul, Shanghai, Shijiazhuang, Taiwan, Tokyo and Warsaw.

LAB MODUS

Lab Modus, founded in 2006 by Kevin Chang, is a collaborative design practice based in Taipei, China with extensive experience ranging from interior to architecture on different scale levels. As an architect and educator Chang is committed to the practice of architecture and interior design as applied research.

Lab Modus intends to define the uniqueness of their performance in the design profession. They believe mutually market demands and client wishes enable their works and they aim for solutions in which their goals and their client's wishes overlap.

ARCHEA ASSOCIATI
architettura - design

Founded in Florence in 1988 by Laura Andreini, Marco Casamonti and Giovanni Polazzi, Archea is today a network of more than 80 architects who work in the firm's seven different branches in Florence, Milan, Rome, Beijing, São Paolo and Dubai. The founding partners were joined in 1999 by Silvia Fabi who coordinates the design activities of Studio Archea in Florence. The most important projects of the firm comprise: the Municipal library of Nembro (Bergamo), the transformation of the former Wine Warehouse of the harbour of Trieste, the social and educatoinal center of Serence, the UBPA B3-2 Pavilion at the World Expo 2010, the Green Energy Laboratory for the Jiao Tong University in Shanghai and the city of ceramics near Li Ling (Hunan), in China, which is nearing completion.

SPARK ✳

SPARK is an award-winning international architectural and design consultancy with proven expertise in architecture, urban design, landscape architecture and interior design. SPARK creates distinctive projects across Asia, Europe and the Middle East. With a dynamic team of over 100 staff spanning 16 nationalities and three continents, SPARK combines the best experience of international and local talent.

Driven by an analytical approach to create architecture that is pragmatic, social and convivial, SPARK works closely with clients to create sustainable architecture that is underpinned by financial viability and the desire to improve the quality of life for all.

With studios in Beijing, London, Shanghai and Singapore, SPARK's award winning projects include Clarke Quay in Singapore, the Shanghai International Cruise Terminal (MIPIM Asia Awards 2011, "Best Mixed-Use development" award), the Starhill Gallery Kuala Lumpur and the Raffles City projects in Ningbo and Beijing.

Established by Chang Ching-Hwa and Kuo Ying-Chao in 1999, Bio architecture Formosana focuses on design sustainability through Green Architecture to create a living environment coexisting with the Nature. Cooperating with other qualified professional involvement, Bio Architecture Formosana carries through it's philosophy, to integrate humans' wellbeing into environment protection through professional execution, delicate design, and critical planning. As the name of the firm has suggested, we've been highly focused on environmental issues in different areas, from reducing urban heat island effect, enhancing bio diversity in man made environment, to integrating renewable resources and eco friendly material in building design. Our works cover a wide range of building types, including library, train station, educational, residential and industrial buildings. In 2007, we establish Bio interior Formosana. It has become a platform that we can implement eco friendly materials and health oriented design in the interior space.

Henn Architekten is an international architectural consultancy with 65 years of expertise in the design and realisation of buildings, masterplans and interior spaces in the fields of culture, administration, teaching and research, development and production as well as urban design.

The office is led by Gunter Henn and eleven partners with offices in Munich, Berlin, Beijing and Shanghai. 350 employees from 25 countries are able to draw upon a wealth of knowledge collected over three generations of building experience in addition to a worldwide network of partners and experts in a variety of disciplines.

This continuity, coupled with progressive design approaches and methods and interdisciplinary research projects, forms the basis for a continual examination of current issues and for a consistent design philosophy. Forms and spaces are no mere objective, they are developed from the processes, demands and cultural contexts of each project. As a general contractor we are able to satisfy this principle at every stage of project planning and implementation.

后记

本书的编写离不开各位设计师和摄影师的帮助，正是有了他们专业而负责的工作态度，才有了本书的顺利出版。参与本书的编写人员有：

COOP HIMMELB(L)AU, Wolf D. Prix, Paul Kath, Wolfgang Reicht, Alexander Ott, Quirin Krumbholz, Eva Wolf, Victoria Coaloa, Heller Manus Architects, Tange Associates, T+E Image Architects, Jeff Walker, Carol Zhang, Johnson Pilton Walker, James Polyhron, Dickson Leung, Adam Rusan, Sophie Blain, Simon Wilson, Andrew Daly, Nicholas Chou, Matthias Knauss, Elisa Nakano, Jan Wesseling, Alex Wilson, Frank Ru, Hannah Ding, Yao Li, Malone Chang, Yu-lin Chen, MAYU architects+, Guei-Shiang Ke, SPARK, John Curran, Lin Ho, Christian Richters, Eric Chan, Arquitectonica, Bernardo Fort Brescia, Laurinda Spear, HYHW Architecture Consulting LTD, Shu He Photography, 9town - studio, Huayang International Design Group, HENN architekten, Gunter Henn, Bartosz Kolonko, RTKL Associates Inc., The Jerde Partnership, Atkins, Ming Lai Architects Inc., lab Modus, Chih-Ming Wu, kokaistudios, Ronald Lu & Partners, blvd. international inc, Du Yun, Bio architecture Formosana, Laura Andreini, Marco Casamonti, Silvia Fabi, Giovanni Polazzi, Studio Archea, JWDA

ACKNOWLEDGEMENTS

We would like to thank everyone involved in the production of this book, especially all the artists, designers, architects and photographers for their kind permission to publish their works. We are also very grateful to many other people whose names do not appear on the credits but who provided assistance and support. We highly appreciate the contribution of images, ideas, and concepts and thank them for allowing their creativity to be shared with readers around the world.